McGraw-Hill
Illustrative Mathematics™
Course 1

mheducation.com/prek-12

Send all inquiries to:
McGraw-Hill Education
STEM Learning Solutions Center
8787 Orion Place
Columbus, OH 43240

ISBN: 978-0-07-689373-7
MHID: 0-07-689373-1

Illustrative Mathematics, Course 1
Student Edition, Volume 2

Printed in the United States of America.

9 10 11 12 LMN 28 27 26 25 24 23 22

Contents in Brief

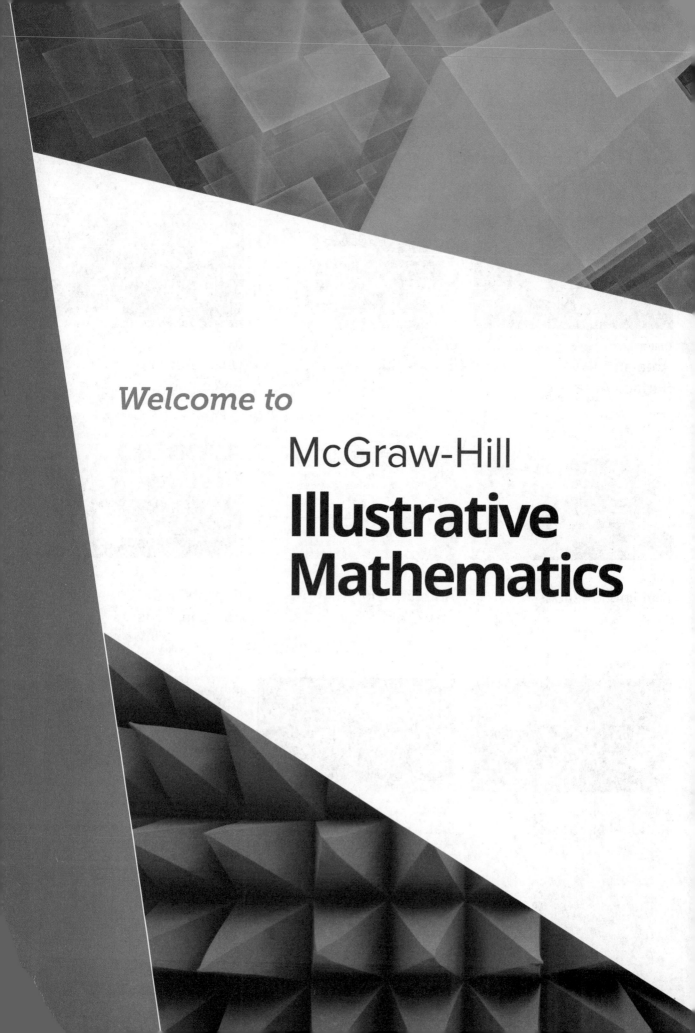

Welcome to

McGraw-Hill

**Illustrative
Mathematics**

Learning math can be an exciting adventure. That's why *Illustrative Mathematics* is designed to provide opportunities for you to learn math by doing math. By engaging in problem-solving activities, you will make and test your own conclusions and try out different strategies as you learn math.

Let's get started!

Unit 1
Area and Surface Area

Unit 2

Introducing Ratios

Unit 3
Unit Rates and Percentages

Unit 4

Dividing Fractions

Unit 5
Arithmetic in Base Ten

Unit 6

Expressions and Equations

Unit 7
Rational Numbers

Negative Numbers and Absolute Value

Inequalities

The Coordinate Plane

Common Factors and Common Multiples

Let's Put It to Work

Unit 8

Data Sets and Distributions

Unit 9

Putting It All Together

Arithmetic in Base Ten

Origami is the Japanese art of folding paper into decorative figures. At the end of this unit, you'll apply what you learned about decimals to create boxes by folding paper.

Topics
- Warming Up to Decimals
- Adding and Subtracting Decimals
- Multiplying Decimals
- Dividing Decimals
- Let's Put It to Work

Unit 5

Arithmetic in Base Ten

Lesson 5-1

Using Decimals in a Shopping Context

NAME _____ DATE _____ PERIOD _____

Learning Goal Let's use what we know about decimals to make shopping decisions.

Warm Up
1.1 Snacks from the Concession Stand

Clare went to a concession stand that sells pretzels for $3.25, drinks for $1.85, and bags of popcorn for $0.99 each. She bought at least one of each item and spent no more than $10.

1. Could Clare have purchased 2 pretzels, 2 drinks, and 2 bags of popcorn? Explain your reasoning.

2. Could she have bought 1 pretzel, 1 drink, and 5 bags of popcorn? Explain your reasoning.

Activity

1.2 Planning a Dinner Party

You are planning a dinner party with a budget of $50 and a menu that consists of 1 main dish, 2 side dishes, and 1 dessert. There will be 8 guests at your party.

Choose your menu items and decide on the quantities to buy so you stay on budget. If you choose meat, fish, or poultry for your main dish, plan to buy at least 0.5 pound per person.

1. The budget is $ _____ per guest.

2. Use the worksheet to record your choices and estimated costs. Then find the estimated total cost and cost per person. See examples in the first two rows.

Item	Quantity Needed	Advertised Price	Estimated Subtotal ($)	Estimated Cost Per Person ($)
Ex. Main Dish: Fish	4 pounds	$6.69 per pound	$4 \cdot 7 = 28$	$28 \div 8 = 3.50$
Ex. Dessert: Cupcakes	8 cupcakes	$2.99 per 6 cupcakes	$2 \cdot 3 = 6$	$6 \div 8 = 0.75$
Main Dish:				
Side Dish 1:				
Side Dish 2:				
Dessert:				
Estimated Total				

NAME _____ DATE _____ PERIOD _____

3. Is your estimated total close to your budget? If so, continue to the next question. If not, revise your menu choices until your estimated total is close to the budget.

4. Calculate the actual costs of the two most expensive items and add them. Show your reasoning.

5. How will you know if your total cost for all menu items will or will not exceed your budget? Is there a way to predict this without adding all the exact costs? Explain your reasoning.

How much would it cost to plant the grass on a football field?
Explain or show your reasoning.

Summary
Using Decimals in a Shopping Context

We often use decimals when dealing with money. In these situations, sometimes we round and make estimates, and other times we calculate the numbers more precisely.

There are many different ways we can add, subtract, multiply, and divide decimals. When we perform these calculations, it is helpful to understand the meanings of the digits in a number and the properties of operations.

We will investigate how these understandings help us work with decimals in upcoming lessons.

NAME _____ DATE _____ PERIOD _____

Practice
Using Decimals in a Shopping Context

1. Mai had $14.50. She spent $4.35 at the snack bar and $5.25 at the arcade. What is the exact amount of money Mai has left?

(A.) $9.60   (C.) $4.90

(B.) $10.60   (D.) $5.90

2. A large cheese pizza costs $7.50. Diego has $40 to spend on pizzas. How many large cheese pizzas can he afford? Explain or show your reasoning.

3. Tickets to a show cost $5.50 for adults and $4.25 for students. A family is purchasing 2 adult tickets and 3 student tickets.

   a. Estimate the total cost.

   b. What is the exact cost?

   c. If the family pays $25, what is the exact amount of change they should receive?

4. Chicken costs $3.20 per pound, and beef costs $4.59 per pound. Answer each question and show your reasoning.

   a. What is the exact cost of 3 pounds of chicken?

   b. What is the exact cost of 3 pound of beef?

   c. How much more does 3 pounds of beef cost than 3 pounds of chicken?

5. Respond to each question. (Lesson 4-16)

a. How many $\frac{1}{5}$-liter glasses can Lin fill with a $1\frac{1}{2}$-liter bottle of water?

b. How many $1\frac{1}{2}$-liter bottles of water does it take to fill a 16-liter jug?

6. Let the side length of each small square on the grid represent 1 unit. (Lesson 4-14)

a. Draw two different triangles, each with base $5\frac{1}{2}$ units and area $19\frac{1}{4}$ units^2.

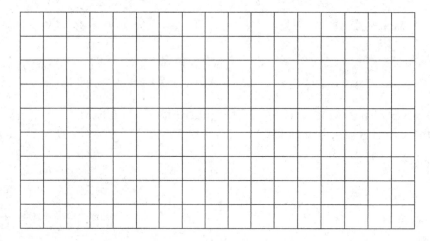

b. Why does each of your triangles have area $19\frac{1}{4}$ units^2? Explain or show your reasoning.

7. Find each quotient. (Lesson 4-10)

a. $\frac{5}{6} \div \frac{1}{6}$

b. $1\frac{1}{6} \div \frac{1}{12}$

c. $\frac{10}{6} \div \frac{1}{24}$

Lesson 5-2

Using Diagrams to Represent Addition and Subtraction

NAME _____ DATE _____ PERIOD _____

Learning Goal Let's represent addition and subtraction of decimals.

Warm Up
2.1 Changing Values

1. Here is a rectangle.

   What number does the rectangle represent
   if each small square represents:

   a. 1

   b. 0.1

   c. 0.01

   d. 0.001

2. Here is a square.

   What number does the square represent
   if each small rectangle represents:

   a. 10

   b. 0.1

   c. 0.00001

You may be familiar with base-ten blocks that represent ones, tens, and hundreds. Here are some diagrams that we will use to represent base-ten units.

- A large square represents 1 one.

- A medium rectangle represents 1 tenth.

- A medium square represents 1 hundredth.

- A small rectangle represents 1 thousandth.

- A small square represents 1 ten-thousandth.

1 one

0.1 tenth

0.01 hundredth

0.001 thousandth

0.0001 ten-thousandth

1. Here is the diagram that Priya drew to represent 0.13. Draw a different diagram that represents 0.13. Explain why both diagrams represent the same number.

2. Here is the diagram that Han drew to represent 0.025. Draw a different diagram that represents 0.025. Explain why both diagrams represent the same number.

NAME _____ DATE _____ PERIOD _____

3. For each number, draw or describe two different diagrams that represent it.

   a. 0.1

   b. 0.02

   c. 0.004

4. Use diagrams of base-ten units to represent each sum. Think about how you could use as few units as possible to represent each number.

   a. 0.03 + 0.05

   b. 0.006 + 0.007

   c. 0.4 + 0.7

Activity

2.3 Finding Sums in Different Ways

1. Here are two ways to calculate the value of 0.26 + 0.07. In the diagram, each rectangle represents 0.1 and each square represents 0.01.

```
      1
    0 . 2 6
  + 0 . 0 7
    0 . 3 3
```

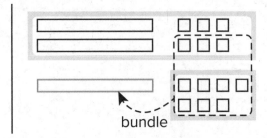

Use what you know about base-ten units and addition to explain:

   a. why ten squares can be "bundled" into a rectangle.

   b. how this "bundling" is represented in the vertical calculation.

2. Find the value of 0.38 + 0.69 by drawing a diagram. Can you find the sum without bundling? Would it be useful to bundle some pieces? Explain your reasoning.

3. Calculate 0.38 + 0.69. Check your calculation against your diagram in the previous question.

4. Find each sum. The larger square represents 1.

   a.
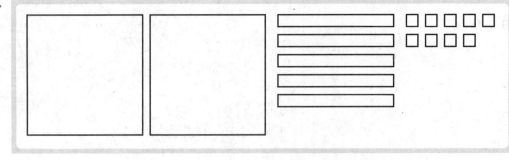

   b.
```
    6 . 0 3
  + 0 . 0 9 8
```

NAME _____ DATE _____ PERIOD _____

Are you ready for more?

A distant, magical land uses jewels for their bartering system. The jewels are valued and ranked in order of their rarity. Each jewel is worth 3 times the jewel immediately below it in the ranking. The ranking is red, orange, yellow, green, blue, indigo, and violet. So a red jewel is worth 3 orange jewels, a green jewel is worth 3 blue jewels, and so on.

1. If you had 500 violet jewels and wanted to trade so that you carried as few jewels as possible, which jewel would you have?

2. Suppose you have 1 orange jewel, 2 yellow jewels, and 1 indigo jewel. If you're given 2 green jewels and 1 yellow jewel, what is the fewest number of jewels that could represent the value of the jewels you have?

Activity

2.4 Representing Subtraction

1. Here are diagrams that represent differences. Removed pieces are marked with Xs. The larger rectangle represents 1 tenth. For each diagram, write a numerical subtraction expression and determine the value of the expression.

Diagram A   **Diagram B**   **Diagram C**

2. Express each subtraction in words.

   a. 0.05 − 0.02

   b. 0.024 − 0.003

   c. 1.26 − 0.14

3. Find each difference by drawing a diagram and by calculating with numbers. Make sure the answers from both methods match. If not, check your diagram and your numerical calculation.

   a. 0.05 − 0.02

   b. 0.024 − 0.003

   c. 1.26 − 0.14

NAME _____ DATE _____ PERIOD _____

Summary
Using Diagrams to Represent Addition and Subtraction

Base-ten diagrams represent collections of base-ten units—tens, ones, tenths, hundredths, etc. We can use them to help us understand sums of decimals.

Suppose we are finding 0.08 + 0.13. Here is a diagram where a square represents 0.01 and a rectangle (made up of ten squares) represents 0.1.

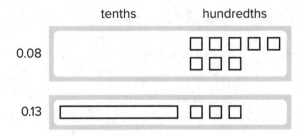

To find the sum, we can "bundle" (or compose) 10 hundredths as 1 tenth.

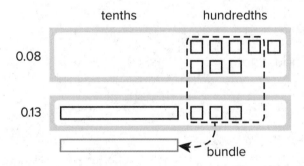

We now have 2 tenths and 1 hundredth, so 0.08 + 0.13 = 0.21.

0.21

We can also use vertical calculation to find 0.08 + 0.13.

Notice how this representation also shows 10 hundredths are bundled (or composed) as 1 tenth.

$$
\begin{array}{r}
1 \\
0.1\ 3 \\
+\ 0.0\ 8 \\
\hline
0.2\ 1
\end{array}
$$

This works for any decimal place. Suppose we are finding 0.008 + 0.013.
Here is a diagram where a small rectangle represents 0.001.

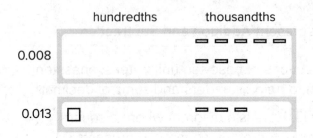

We can "bundle" (or compose) 10 thousandths as 1 hundredth.

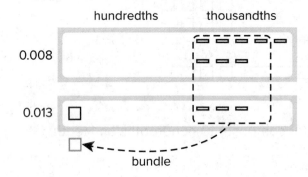

bundle

The sum is 2 hundredths and 1 thousandth

0.021

Here is a vertical calculation of 0.008 + 0.013.

$$
\begin{array}{r}
{\scriptstyle 1} \\
0.0\,1\,3 \\
+\ \ 0.0\,0\,8 \\
\hline
0.0\,2\,1
\end{array}
$$

NAME _____ DATE _____ PERIOD _____

Practice
Using Diagrams to Represent Addition and Subtraction

1. Use the given key to answer the questions.

   a. What number does this diagram represent?

   b. Draw a diagram that represents 0.216.

   c. Draw a diagram that represents 0.304.

2. Here are diagrams that represent 0.137 and 0.284.

   tenths   hundredths   thousandths

   a. Use the diagram to find the value of 0.137 + 0.284. Explain your reasoning.

b. Calculate the sum vertically.

$$
\begin{array}{r}
0.\ 1\ 3\ 7 \\
+\ \ 0.\ 2\ 8\ 4 \\
\hline
\end{array}
$$

c. How was your reasoning about 0.137 + 0.284 the same with the two methods? How was it different?

3. For the first two problems, circle the vertical calculation where digits of the same kind are lined up. Then, finish the calculation and find the sum. For the last two problems, find the sum using vertical calculation.

a. 3.25 + 1

$$
\begin{array}{r}
3.2\ 5 \\
+\ \ \ \ \ 1.0 \\
\hline
\end{array}
\qquad
\begin{array}{r}
3.2\ 5 \\
+1.0 \\
\hline
\end{array}
\qquad
\begin{array}{r}
3.2\ 5 \\
+\ \ \ \ \ \ \ \ 1 \\
\hline
\end{array}
$$

b. 0.5 + 1.15

$$
\begin{array}{r}
0.5 \\
+\ 1.1\ 5 \\
\hline
\end{array}
\qquad
\begin{array}{r}
0.5 \\
+\ 1.1\ 5 \\
\hline
\end{array}
\qquad
\begin{array}{r}
0.5\ 0 \\
+\ 1.1\ 5\ 0 \\
\hline
\end{array}
$$

c. 10.6 + 1.7

d. 123 + 0.2

4. Andre has been practicing his math facts. He can now complete 135 multiplication facts in 90 seconds. **(Lesson 2-9)**

   a. If Andre is answering questions at a constant rate, how many facts can he answer per second?

   b. Noah also works at a constant rate, and he can complete 75 facts in 1 minute. Who is working faster? Explain or show your reasoning.

Lesson 5-3

Adding and Subtracting Decimals with Few Non-Zero Digits

NAME _____ DATE _____ PERIOD _____

Learning Goal Let's add and subtract decimals.

Warm Up
3.1 Do the Zeros Matter?

1. Evaluate mentally: $1.009 + 0.391$

2. Decide if each equation is true or false. Be prepared to explain your reasoning.

   a. $34.56000 = 34.56$ **b.** $25 = 25.0$ **c.** $2.405 = 2.45$

Activity
3.2 Calculating Sums

1. Andre and Jada drew base-ten diagrams to represent $0.007 + 0.004$. Andre drew 11 small rectangles. Jada drew only two figures: a square and a small rectangle.

   Andre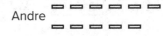

   Jada

   a. If both students represented the sum correctly, what value does each small rectangle represent? What value does each square represent?

   b. Draw or describe a diagram that could represent the sum $0.008 + 0.07$.

2. Here are two calculations of 0.2 + 0.05. Which is correct? Explain why one is correct and the other is incorrect.

$$
\begin{array}{r}
0.2 \\
+\ 0.0\,5 \\
\hline
0.2\,5
\end{array}
\qquad\qquad
\begin{array}{r}
0.2 \\
+\ 0.0\,5 \\
\hline
0.0\,7
\end{array}
$$

3. Compute each sum. If you get stuck, consider drawing base-ten diagrams to help you.

a.
$$
\begin{array}{r}
0.1\,1 \\
+\ 0.0\,0\,5 \\
\hline
\end{array}
$$

b. 0.209 + 0.01

c. 10.2 + 1.1456

NAME _____ DATE _____ PERIOD _____

 Activity

3.3 Subtracting Decimals of Different Lengths

Diego and Noah drew different diagrams to represent 0.4 − 0.03.
Each rectangle represents 0.1. Each square represents 0.01.

Diego started by drawing 4 rectangles to represent 0.4. He then replaced 1 rectangle with 10 squares and crossed out 3 squares to represent subtraction of 0.03, leaving 3 rectangles and 7 squares in his diagram.

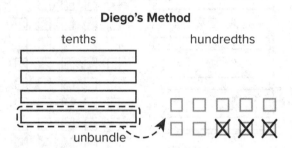

Noah started by drawing 4 rectangles to represent 0.4. He then crossed out 3 rectangles to represent the subtraction, leaving 1 rectangle in his diagram.

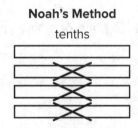

1. Do you agree that either diagram correctly represents 0.4 − 0.03? Discuss your reasoning with a partner.

2. To represent 0.4 − 0.03, Elena drew another diagram. She also started by drawing 4 rectangles. She then replaced all 4 rectangles with 40 squares and crossed out 3 squares to represent subtraction of 0.03, leaving 37 squares in her diagram. Is her diagram correct? Discuss your reasoning with a partner.

Elena's Method

3. Find each difference. Explain or show your reasoning.

   a. 0.3 − 0.05

   b. 2.1 − 0.4

   c. 1.03 − 0.06

   d. 0.02 − 0.007

NAME _____ DATE _____ PERIOD _____

Are you ready for more?

A distant, magical land uses jewels for their bartering system. The jewels are valued and ranked in order of their rarity. Each jewel is worth 3 times the jewel immediately below it in the ranking. The ranking is red, orange, yellow, green, blue, indigo, and violet. So a red jewel is worth 3 orange jewels, a green jewel is worth 3 blue jewels, and so on.

At the Auld Shoppe, a shopper buys items that are worth 2 yellow jewels, 2 green jewels, 2 blue jewels, and 1 indigo jewel. If they came into the store with 1 red jewel, 1 yellow jewel, 2 green jewels, 1 blue jewel, and 2 violet jewels, what jewels do they leave with? Assume the shopkeeper gives them their change using as few jewels as possible.

Base-ten diagrams can help us understand subtraction as well.

Suppose we are finding 0.23 − 0.07. Here is a diagram showing 0.23, or 2 tenths and 3 hundredths.

Subtracting 7 hundredths means removing 7 small squares, but we do not have enough to remove. Because 1 tenth is equal to 10 hundredths, we can "unbundle" (or decompose) one of the tenths (1 rectangle) into 10 hundredths (10 small squares).

We now have 1 tenth and 13 hundredths, from which we can remove 7 hundredths.

We have 1 tenth and 6 hundredths remaining, so 0.23 − 0.07 = 0.16.

Here is a vertical calculation of 0.23 − 0.07.

$$
\begin{array}{r}
\overset{1}{}\overset{13}{} \\
0.\cancel{2}\,\cancel{3} \\
-\ 0.0\ 7 \\
\hline
0.1\ 6
\end{array}
$$

Notice how this representation also shows a tenth is unbundled (or decomposed) into 10 hundredths in order to subtract 7 hundredths.

NAME _____ DATE _____ PERIOD _____

This works for any decimal place. Suppose we are finding 0.023 − 0.007. Here is a diagram showing 0.023.

We want to remove 7 thousandths (7 small rectangles). We can "unbundle" (or decompose) one of the hundredths into 10 thousandths.

Now we can remove 7 thousandths.

We have 1 hundredth and 6 thousandths remaining, so 0.023 − 0.007 = 0.016.

Here is a vertical calculation of 0.023 − 0.007.

$$\begin{array}{r} \overset{\scriptstyle 1 \quad 13}{0.0\,\cancel{2}\,\cancel{3}} \\ -\ 0.0\,0\,7 \\ \hline 0.0\,1\,6 \end{array}$$

Practice
Adding and Subtracting Decimals with Few Non-Zero Digits

1. Here is a base-ten diagram that represents 1.13. Use the diagram to find 1.13 − 0.46. Explain or show your reasoning.

2. Compute the following sums. If you get stuck, consider drawing base-ten diagrams.

   a. 0.027 + 0.004

   b. 0.203 + 0.01

   c. 1.2 + 0.145

NAME _____ DATE _____ PERIOD _____

3. A student said we cannot subtract 1.97 from 20 because 1.97 has two decimal digits and 20 has none. Do you agree with him? Explain or show your reasoning.

4. Decide which calculation shows the correct way to find 0.3 − 0.006 and explain your reasoning.

(A.)
```
    0 . 3
−  0 . 0 0 6
─────────────
  0 . 3 0 6
```

(B.)
```
       0 . 3
−   0 . 0 0 6
─────────────
   0 . 0 9 7
```

(C.)
```
    0 . 3 0
−  0 . 0 0 6
─────────────
  0 . 0 2 4
```

(D.)
```
    0 . 3 0 0
−  0 . 0 0 6
─────────────
   0 . 2 9 4
```

5. Complete the calculations so that each shows the correct difference.

a.
```
   1  4  2 . 6
−         1 . 4
──────────────
  [  ][  ][  ] . 2
```

b.
```
   3  8 . 6  0
−      6 . 7  5
──────────────
  [  ][  ] . [  ] 5
```

c.
```
   2  4  1 . 7  6
−         2 . 1  8
──────────────
  [  ][  ][  ] . [  ] 8
```

6. The school store sells pencils for $0.30 each, hats for $14.50 each, and binders for $3.20 each. Elena would like to buy 3 pencils, a hat, and 2 binders. She estimated that the cost will be less than $20. (Lesson 5-1)

   a. Do you agree with her estimate? Explain your reasoning.

   b. Estimate the number of pencils she could buy with $5. Explain or show your reasoning.

7. A rectangular prism measures $7\frac{1}{2}$ cm by 12 cm by $15\frac{1}{2}$ cm. (Lesson 4-15)

   a. Calculate the number of cubes with edge length $\frac{1}{2}$ cm that fit in this prism.

   b. What is the volume of the prism in cm³? Show your reasoning. If you are stuck, think about how many cubes with $\frac{1}{2}$-cm edge lengths fit into 1 cm³.

8. At a constant speed, a car travels 75 miles in 60 minutes. How far does the car travel in 18 minutes? If you get stuck, consider using the table.
(Lesson 2-12)

Minutes	Distance in Miles
60	75
6	
18	

Lesson 5-4

Adding and Subtracting Decimals with Many Non-Zero Digits

NAME _____ DATE _____ PERIOD _____

Learning Goal Let's practice adding and subtracting decimals.

 ## Warm Up
4.1 The Cost of a Photo Print

1. Here are three ways to write a subtraction calculation. What do you notice? What do you wonder?

$$
\begin{array}{r} 5 \\ -0.17 \\ \hline \end{array}
\qquad
\begin{array}{r} 5 \\ -0.17 \\ \hline \end{array}
\qquad
\begin{array}{r} 5 \\ -0.17 \\ \hline \end{array}
$$

2. Clare bought a photo for 17 cents and paid with a $5 bill. Look at the previous question. Which way of writing the numbers could Clare use to find the change she should receive? Be prepared to explain how you know.

3. Find the amount of change that Clare should receive. Show your reasoning, and be prepared to explain how you calculate the difference of 0.17 and 5.

Activity

4.2 Decimals All Around

1. Find the value of each expression. Show your reasoning.

   a. $11.3 - 9.5$

   b. $318.8 - 94.63$

   c. $0.02 - 0.0116$

2. Discuss with a partner:

   a. Which method or methods did you use in the previous question? Why?

   b. In what ways were your methods effective? Was there an expression for which your methods did not work as well as expected?

3. Lin's grandmother ordered needles that were 0.3125 inches long to administer her medication, but the pharmacist sent her needles that were 0.6875 inches long. How much longer were these needles than the ones she ordered? Show your reasoning.

4. There is 0.162 liter of water in a 1-liter bottle. How much more water should be put in the bottle so it contains exactly 1 liter? Show your reasoning.

5. One micrometer is 1 millionth of a meter. A red blood cell is about 7.5 micrometers in diameter. A coarse grain of sand is about 70 micrometers in diameter. Find the difference between the two diameters in *meters*. Show your reasoning.

NAME _____ DATE _____ PERIOD _____

Activity
4.3 Missing Numbers

Write the missing digits in each calculation so that the value of each sum or difference is correct. Be prepared to explain your reasoning.

1.
```
   0. 4  0  4
+ [  |  |  |  ]
─────────────
   1
```

2.
```
      9. 8  7  6  5
+ [  |  |  |  |  ]
──────────────────
      1  0
```

3.
```
   0. 7
− [  |  |  ]
────────────
   0. 0  1  2
```

4.
```
   7
− [  |  |  |  ]
────────────────
   3. 4  5  6  7
```

5.
```
   7  0
− [  |  |  |  ]
────────────────
   0. 0  0  8  9
```

Are you ready for more?

In a cryptarithmetic puzzle, the digits 0-9 are represented using the first 10 letters of the alphabet. Use your understanding of decimal addition to determine which digits go with the letters A, B, C, D, E, F, G, H, I, and J. How many possibilities can you find?

```
  I H F . I J
+ J I I . F I
─────────────
  E J I . I E
```

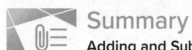

Summary

Adding and Subtracting Decimals with Many Non-Zero Digits

Base-ten diagrams work best for representing subtraction of numbers with few non-zero digits, such as 0.16 − 0.09. For numbers with many non-zero digits, such as 0.25103 − 0.04671, it would take a long time to draw the base-ten diagram. With vertical calculations, we can find this difference efficiently.

Thinking about base-ten diagrams can help us make sense of this calculation.

The thousandth in 0.25103 is unbundled (or decomposed) to make 10 ten-thousandths so that we can subtract 7 ten-thousandths. Similarly, one of the hundredths in 0.25103 is unbundled (or decomposed) to make 10 thousandths.

$$
\begin{array}{r}
\overset{\overset{\displaystyle 10}{\overset{\displaystyle 4\ \cancel{0}10}{}}}{0.25\cancel{1}\cancel{0}3} \\
-\ 0.04671 \\
\hline
0.20432
\end{array}
$$

NAME _____ DATE _____ PERIOD _____

 Practice

Adding and Subtracting Decimals with Many Non-Zero Digits

1. For each subtraction problem, circle the correct calculation.

   a. 7.2 − 3.67

7.2	0 7.2	7.2 0
−3.6 7	−3.6 7	−3.6 7
3.0 5	3.0 5	3.5 3

   b. 16 − 1.4

1 6	1 6.0	1 6.0
− 1.4	−1.4 0	− 1.4
0.2	0.2 0	14.6

2. Explain how you could find the difference of 1 and 0.1978.

3. A bag of chocolates is labeled to contain 0.384 pound of chocolates. The actual weight of the chocolates is 0.3798 pound.

    a. Are the chocolates heavier or lighter than the weight stated on the label? Explain how you know.

    b. How much heavier or lighter are the chocolates than stated on the label? Show your reasoning.

NAME _____ DATE _____ PERIOD _____

4. Complete the calculations so that each shows the correct sum.

a.
```
   1 . 0   3   6
 + [  ][  ][  ][  ]
 ─────────────────
   4
```

b.
```
   0 . 7   3   8
 + [  ][  ][  ][  ]
 ─────────────────
   1
```

c.
```
   0 . 5   1   3   7
 + [  ][  ][  ][  ][  ]
 ──────────────────────
   1
```

5. A shipping company is loading cube-shaped crates into a larger cube-shaped container. The smaller cubes have side lengths of $2\frac{1}{2}$ feet, and the larger shipping container has side lengths of 10 feet. How many crates will fit in the large shipping container? Explain your reasoning. **(Lesson 4-14)**

6. For every 9 customers, the chef prepares 2 loaves of bread. (Lesson 2-13)

a. Here is a double number line showing varying numbers of customers and the loaves prepared. Complete the missing information.

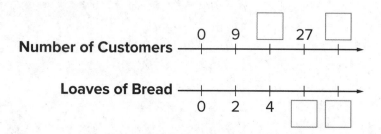

b. The same information is shown on a table. Complete the missing information.

Customers	Loaves
9	2
	4
27	
	14
1	

c. Use either representation to answer these questions.

• How many loaves are needed for 63 customers?

• How many customers are there if the chef prepares 20 loaves?

• How much of a loaf is prepared for each customer?

Lesson 5-5

Decimal Points in Products

NAME _____ DATE _____ PERIOD _____

Learning Goal Let's look at products that are decimals.

 Warm Up
5.1 Multiplying by 10

1. In which equation is the value of x the largest?

$x \cdot 10 = 810$ $x \cdot 10 = 81$ $x \cdot 10 = 8.1$ $x \cdot 10 = 0.81$

2. How many times the size of 0.81 is 810?

 Activity
5.2 Fractionally Speaking: Powers of Ten

Work with a partner. One person solves the problems labeled "Partner A" and the other person solves those labeled "Partner B." Then compare your results.

1. Find each product or quotient. Be prepared to explain your reasoning.

Partner A

a. $250 \cdot \frac{1}{10}$

b. $250 \cdot \frac{1}{100}$

c. $48 \div 10$

d. $48 \div 100$

Partner B

a. $250 \div 10$

b. $250 \div 100$

c. $48 \cdot \frac{1}{10}$

d. $48 \cdot \frac{1}{100}$

2. Use your work in the previous problems to find 720 · (0.1) and 720 · (0.01). Explain your reasoning.

   Pause here for a class discussion.

3. Find each product. Show your reasoning.

   a. 36 · (0.1)

   b. (24.5) · (0.1)

   c. (1.8) · (0.1)

   d. 54 · (0.01)

   e. (9.2) · (0.01)

4. Jada says: "If you multiply a number by 0.001, the decimal point of the number moves three places to the left." Do you agree with her? Explain your reasoning.

 Activity

5.3 Fractionally Speaking: Multiples of Powers of Ten

1. Select **all** expressions that are equivalent to (0.6) · (0.5). Be prepared to explain your reasoning.

   (A.) 6 · (0.1) · 5 · (0.1)

   (B.) 6 · (0.01) · 5 · (0.1)

   (C.) $6 \cdot \frac{1}{10} \cdot 5 \cdot \frac{1}{10}$

   (D.) $6 \cdot \frac{1}{1,000} \cdot 5 \cdot \frac{1}{100}$

   (E.) 6 · (0.001) · 5 · (0.01)

   (F.) $6 \cdot 5 \cdot \frac{1}{10} \cdot \frac{1}{10}$

   (G.) $\frac{6}{10} \cdot \frac{5}{10}$

NAME _____ DATE _____ PERIOD _____

2. Find the value of (0.6) • (0.5). Show your reasoning.

3. Find the value of each product by writing and reasoning with an equivalent expression with fractions.

   a. (0.3) • (0.02)

   b. (0.7) • (0.05)

Are you ready for more?

Ancient Romans used the letter I for 1, V for 5, X for 10, L for 50, C for 100, D for 500, and M for 1,000.

Write a problem involving merchants at an agora, an open-air market, that uses multiplication of numbers written with Roman numerals.

Summary
Decimal Points in Products

We can use fractions like $\frac{1}{10}$ and $\frac{1}{100}$ to reason about the location of the decimal point in a product of two decimals.

Let's take $24 \cdot (0.1)$ as an example. There are several ways to find the product:

- We can interpret it as 24 groups of 1 tenth (or 24 tenths), which is 2.4.
- We can think of it as $24 \cdot \frac{1}{10}$, which is equal to $\frac{24}{10}$ (and also equal to 2.4).
- Multiplying by $\frac{1}{10}$ has the same result as dividing by 10, so we can also think of the product as $24 \div 10$, which is equal to 2.4.

Similarly, we can think of $(0.7) \cdot (0.09)$ as 7 tenths times 9 hundredths, and write:

$$\left(7 \cdot \frac{1}{10}\right) \cdot \left(9 \cdot \frac{1}{100}\right)$$

We can rearrange whole numbers and fractions:

$$(7 \cdot 9) \cdot \left(\frac{1}{10} \cdot \frac{1}{100}\right) = 63 \cdot \frac{1}{1,000} = \frac{63}{1,000}$$

This tells us that $(0.7) \cdot (0.09) = 0.063$.

Here is another example.

To find $(1.5) \cdot (0.43)$, we can think of 1.5 as 15 tenths and 0.43 as 43 hundredths. We can write the tenths and hundredths as fractions and rearrange the factors.

$$\left(15 \cdot \frac{1}{10}\right) \cdot \left(43 \cdot \frac{1}{100}\right) = 15 \cdot 43 \cdot \frac{1}{1,000}$$

Multiplying 15 and 43 gives us 645, and multiplying $\frac{1}{10}$ and $\frac{1}{100}$ gives us $\frac{1}{1,000}$. So $(1.5) \cdot (0.43)$ is $645 \cdot \frac{1}{1,000}$, which is 0.645.

NAME _____ DATE _____ PERIOD _____

Practice
Decimal Points in Products

1. Respond to each of the following.

   a. Find the product of each number and $\frac{1}{100}$.

   122.1   11.8   1,350.1   1.704

   b. What happens to the decimal point of the original number when you multiply it by $\frac{1}{100}$? Why do you think that is? Explain your reasoning.

2. Which expression has the same value as $(0.06) \cdot (0.154)$? Select **all** that apply.

   A. $6 \cdot \frac{1}{100} \cdot 154 \cdot \frac{1}{1,000}$

   B. $6 \cdot 154 \cdot \frac{1}{100,000}$

   C. $6 \cdot (0.1) \cdot 154 \cdot (0.01)$

   D. $6 \cdot 154 \cdot (0.00001)$

   E. 0.00924

3. Calculate the value of each expression by writing the decimal factors as fractions, then writing their product as a decimal. Show your reasoning.

   a. $(0.01) \cdot (0.02)$ b. $(0.3) \cdot (0.2)$

   c. $(1.2) \cdot 5$ d. $(0.9) \cdot (1.1)$

   e. $(1.5) \cdot 2$

4. Write three numerical expressions that are equivalent to $(0.0004) \cdot (0.005)$.

5. Calculate each sum. (Lesson 5-3)

a. $33.1 + 1.95$

b. $1.075 + 27.105$

c. $0.401 + 9.28$

6. Calculate each difference. Show your reasoning. (Lesson 5-4)

a. $13.2 - 1.78$

b. $23.11 - 0.376$

c. $0.9 - 0.245$

7. On the grid, draw a quadrilateral *that is not a rectangle* that has an area of 18 square units. Show how you know the area is 18 square units. (Lesson 1-3)

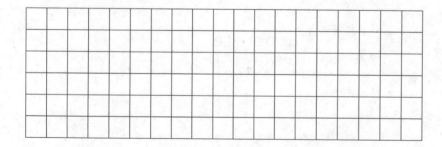

Lesson 5-6

Methods for Multiplying Decimals

NAME _____ DATE _____ PERIOD _____

Learning Goal Let's look at some ways we can represent multiplication of decimals.

Warm Up
6.1 Equivalent Expressions

Write as many expressions as you can think of that are equal to 0.6.
Do not use addition or subtraction.

Activity
6.2 Using Properties of Numbers to Reason about Multiplication

Elena and Noah used different methods to compute $(0.23) \cdot (1.5)$.
Both calculations were correct.

Elena's Method	**Noah's Method**
$(0.23) \cdot 100 = 23$	$0.23 = \dfrac{23}{100}$
$(1.5) \cdot 10 = 15$	$1.5 = \dfrac{15}{10}$
$23 \cdot 15 = 345$	$\dfrac{23}{100} \cdot \dfrac{15}{10} = \dfrac{345}{1,000}$
$345 \div 1,000 = 0.345$	$\dfrac{345}{1,000} = 0.345$

1. Analyze the two methods, then discuss these questions with your partner.

   a. Which method makes more sense to you? Why?

   b. What might Elena do to compute $(0.16) \cdot (0.03)$? What might Noah do to compute $(0.16) \cdot (0.03)$? Will the two methods result in the same value?

2. Compute each product using the equation 21 • 47 = 987 and what you know about fractions, decimals, and place value. Explain or show your reasoning.

   a. (2.1) • (4.7)

   b. 21 • (0.047)

   c. (0.021) • (4.7)

 Activity

6.3 Using Area Diagrams to Reason about Multiplication

1. In the diagram, the side length of each square is 0.1 unit.

   a. Explain why the area of each square is *not* 0.1 square unit.

   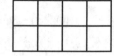

   b. How can you use the area of each square to find the area of the rectangle? Explain or show your reasoning.

   c. Explain how the diagram shows that the equation (0.4) • (0.2) = 0.08 is true.

2. Label the squares with their side lengths so the area of this rectangle represents 40 • 20.

   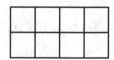

   a. What is the area of each square?

   b. Use the squares to help you find 40 • 20. Explain or show your reasoning.

3. Label the squares with their side lengths so the area of this rectangle represents (0.04) • (0.02).

   Next, use the diagram to help you find (0.04) • (0.02). Explain or show your reasoning.

Summary
Methods for Multiplying Decimals

Here are three other ways to calculate a product of two decimals such as $(0.04) \cdot (0.07)$.

- First, we can multiply each decimal by the same power of 10 to obtain whole-number factors.

$(0.04) \cdot 100 = 4$

$(0.07) \cdot 100 = 7$

Because we multiplied both 0.04 and 0.07 by 100 to get 4 and 7, the product 28 is $(100 \cdot 100)$ times the original product, so we need to divide 28 by 10,000.

$4 \cdot 7 = 28$

$28 \div 10,000 = 0.0028$

- Second, we can write each decimal as a fraction, $0.04 = \dfrac{4}{100}$ and $0.07 = \dfrac{7}{100}$, and multiply them.

$\dfrac{4}{100} \cdot \dfrac{7}{100} = \dfrac{28}{10,000}$

$= 0.0028$

- Third, we can use an area model. The product $(0.04) \cdot (0.07)$ can be thought of as the area of a rectangle with side lengths of 0.04 unit and 0.07 unit.

In this diagram, each small square is 0.01 unit by 0.01 unit. The area of each square, in square units, is therefore $\left(\dfrac{1}{100} \cdot \dfrac{1}{100}\right)$, which is $\dfrac{1}{10,000}$.

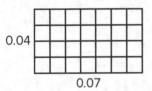

Because the rectangle is composed of 28 small squares, the area of the rectangle, in square units, must be:

$$28 \cdot \dfrac{1}{10,000} = \dfrac{28}{10,000}$$
$$= 0.0028$$

All three calculations show that $(0.04) \cdot (0.07) = 0.0028$.

Methods for Multiplying Decimals

1. Find each product. Show your reasoning.

   a. $(1.2) \cdot (0.11)$

   b. $(0.34) \cdot (0.02)$

   c. $120 \cdot (0.002)$

2. You can use a rectangle to represent $(0.3) \cdot (0.5)$.

   a. What must the side length of each square represent for the rectangle to correctly represent $(0.3) \cdot (0.5)$?

   b. What area is represented by each square?

   c. What is $(0.3) \cdot (0.5)$? Show your reasoning.

NAME _____ DATE _____ PERIOD _____

3. One gallon of gasoline in Buffalo, New York costs $2.29. In Toronto, Canada, one liter of gasoline costs $0.91. There are 3.8 liters in one gallon.

    a. How much does one gallon of gas cost in Toronto? Round your answer to the nearest cent.

    b. Is the cost of gas greater in Buffalo or in Toronto? How much greater?

4. Calculate each sum or difference. **(Lesson 5-2)**

    a. $10.3 + 3.7$

    b. $20.99 - 4.97$

    c. $15.99 + 23.51$

    d. $1.893 - 0.353$

5. Find the value of $\frac{49}{50} \div \frac{7}{6}$ using any method. **(Lesson 4-11)**

6. Find the area of the shaded region. All angles are right angles. Show your reasoning. (Lesson 1-1)

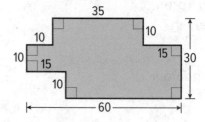

7. a. Priya finds (1.05) • (2.8) by calculating 105 • 28, then moving the decimal point three places to the left. Why does Priya's method make sense?

b. Use Priya's method to calculate (1.05) • (2.8). You can use the fact that 105 • 28 = 2,940.

c. Use Priya's method to calculate (0.0015) • (0.024).

Lesson 5-7

Using Diagrams to Represent Multiplication

NAME _____ DATE _____ PERIOD _____

Learning Goal Let's use area diagrams to find products.

 ## Warm Up
7.1 Estimate the Product

For each of the following products, choose the best estimate of its value.
Be prepared to explain your reasoning.

1. (6.8) · (2.3)

- 1.40
- 14
- 140

2. 74 · (8.1)

- 5.6
- 56
- 560

3. 166 · (0.09)

- 1.66
- 16.6
- 166

4. (3.4) · (1.9)

- 6.5
- 65
- 650

Activity

7.2 Connecting Area Diagrams to Calculations with Whole Numbers

1. Here are three ways of finding the area of a rectangle that is 24 units by 13 units.

Diagram 1

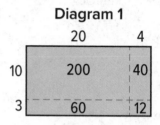

Diagram 2

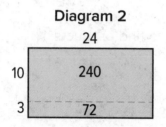

Diagram 3

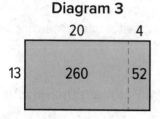

a. What do the diagrams have in common? How are they the same?

b. How are the diagrams different?

c. If you were to find the area of a rectangle that is 37 units by 19 units, which of the three ways of decomposing the rectangle would you use? Why?

2. You may be familiar with different ways to write multiplication calculations. Here are two ways to calculate 24 times 13.

Calculation A

```
      2 4
  ×   1 3
      1 2 ⎤
      6 0 ⎥ partial
      4 0 ⎥ products
  + 2 0 0 ⎦
    3 1 2
```

Calculation B

```
        2 4
    ×   1 3
        7 2
  +   2 4 0
      3 1 2
```

a. In Calculation A, how are each of the partial products obtained? For instance, where does the 12 come from?

b. In Calculation B, how are the 72 and 240 obtained?

NAME _____ DATE _____ PERIOD _____

  c. Look at the diagrams in the first question. Which diagram corresponds
     to Calculation A? Which one corresponds to Calculation B?

  d. How are the partial products in Calculation A and the 72 and 240 in
     Calculation B related to the numbers in the diagrams?

3. Use the two following methods to find the product of 18 and 14.

  a. Calculate
     numerically

```
      1 8
  ×   1 4
```

  b. Here is a rectangle that is
     18 units by 14 units. Find its area,
     in square units by decomposing
     it. Show your reasoning.

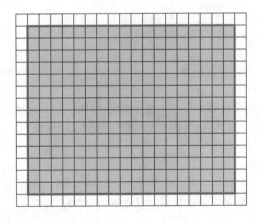

4. Compare the values of 18 · 14 that you obtained using the two methods.
   If they are not the same, check your work.

7.3 Connecting Area Diagrams to Calculations with Decimals

1. You can use area diagrams to represent products of decimals. Here is an area diagram that represents (2.4) • (1.3).

   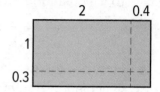

   a. Find the region that represents (0.4) • (0.3). Label it with its area of 0.12.

   b. Label each of the other regions with their respective areas.

   c. Find the value of (2.4) • (1.3). Show your reasoning.

2. Here are two ways of calculating (2.4) • (1.3). Analyze the calculations and discuss the following questions with a partner.

   Calculation A

   ```
         2 . 4
   ×     1 . 3
       0 . 1  2  ⎤
       0 . 6     ⎥ partial
       0 . 4     ⎥ products
   +   2         ⎦
       3 . 1  2
   ```

   Calculation B

   ```
         2 . 4
   ×     1 . 3
       0 . 7  2
   +   2 . 4
       3 . 1  2
   ```

   a. In Calculation A, where does the 0.12 and other partial products come from? In Calculation B, where do the 0.72 and 2.4 come from?

   b. In each calculation, why are the numbers below the horizontal line aligned vertically the way they are?

3. Find the product of (3.1) • (1.5) by drawing and labeling an area diagram. Show your reasoning.

4. Show how to calculate (3.1) • (1.5) using numbers without a diagram. Be prepared to explain your reasoning. If you are stuck, use the examples in a previous question to help you.

NAME _____ DATE _____ PERIOD _____

How many hectares is the property of your school? How many morgens is that?

Activity

7.4 Using the Partial Products Method

1. Label the area diagram to represent $(2.5) \cdot (1.2)$ and to find that product.

   a. Decompose each number into its base-ten units (ones, tenths, etc.) and write them in the boxes on each side of the rectangle.

   b. Label Regions A, B, C, and D with their areas. Show your reasoning.

   c. Find the product that the area diagram represents. Show your reasoning.

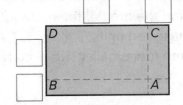

2. Here are two ways to calculate $(2.5) \cdot (1.2)$. Each number with a box gives the area of one or more regions in the area diagram.

   Calculation A

   $$
   \begin{array}{r}
   2.5 \\
   \times \quad 1.2 \\
   \hline
   0.1 \\
   0.4 \\
   0.5 \\
   +\ 2.0 \\
   \hline
   3.0\,0
   \end{array}
   $$

   Calculation B

   $$
   \begin{array}{r}
   2.5 \\
   \times \quad 1.2 \\
   \hline
   0.5 \\
   +\ 2.5 \\
   \hline
   3.0\,0
   \end{array}
   $$

   a. In the boxes next to each number, write the letter(s) of the corresponding region(s).

   b. In Calculation B, which two numbers are being multiplied to obtain 0.5? Which numbers are being multiplied to obtain 2.5?

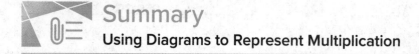
Suppose that we want to calculate the product of two numbers that are written in base ten. To explain how, we can use what we know about base-ten numbers and areas of rectangles.

Here is a diagram of a rectangle with side lengths 3.4 units and 1.2 units.

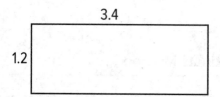

Its area, in square units, is the product $(3.4) \cdot (1.2)$. To calculate this product and find the area of the rectangle, we can decompose each side length into its base-ten units, $3.4 = 3 + 0.4$ and $1.2 = 1 + 0.2$, decomposing the rectangle into four smaller sub-rectangles.

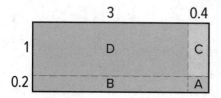

We can rewrite the product and expand it twice:

$$
\begin{aligned}
(3.4) \cdot (1.2) &= (3 + 0.4) \cdot (1 + 0.2) \\
&= (3 + 0.4) \cdot 1 + (3 + 0.4) \cdot 0.2 \\
&= 3 \cdot 1 + 3 \cdot (0.2) + (0.4) \cdot 1 + (0.4) \cdot (0.2)
\end{aligned}
$$

In the last expression, each of the four terms is called a partial product. Each partial product gives the area of a sub-rectangle in the diagram. The sum of the four partial products gives the area of the entire rectangle.

NAME _____ DATE _____ PERIOD _____

We can show the horizontal calculations as two vertical calculations.

$$
\begin{array}{r}
3.4 \\
\times \quad 1.2 \\
\hline
{\scriptstyle 1} \\
0.08 \\
0.6 \\
0.4 \\
+ \quad 3 \\
\hline
4.08
\end{array}
\quad
\begin{array}{cc}
\text{A} \\
\text{B} \\
\text{C} \\
\text{D}
\end{array}
\qquad
\begin{array}{r}
3.4 \\
\times \quad 1.2 \\
\hline
{\scriptstyle 1} \\
0.68 \\
+ \quad 3.4 \\
\hline
4.08
\end{array}
\quad
\begin{array}{c}
\text{A} + \text{B} \\
\text{C} + \text{D}
\end{array}
$$

The vertical calculation on the left is an example of the partial products method. It shows the values of each partial product and the letter of the corresponding sub-rectangle. Each partial product gives an area:

- A is 0.2 unit by 0.4 unit, so its area is 0.08 square unit.

- B is 3 unit by 0.2 unit, so its area is 0.6 square unit.

- C is 0.4 unit by 1 unit, so its area is 0.4 square unit.

- D is 3 units by 1 unit, so its area is 3 square units.

- The sum of the partial products is $0.08 + 0.6 + 0.4 + 3$, so the area of the rectangle is 4.08 square units.

The calculation on the right shows the values of two products. Each value gives a combined area of two sub-rectangles:

- The combined regions of A and B have an area of 0.68 square units; 0.68 is the value of $(3 + 0.4) \cdot 0.2$.

- The combined regions of C and D have an area of 3.4 square units; 3.4 is the value of $(3 + 0.4) \cdot 1$.

- The sum of the values of two products is $0.68 + 3.4$, so the area of the rectangle is 4.08 square units.

Practice

Using Diagrams to Represent Multiplication

1. Here is a rectangle that has been partitioned into four smaller rectangles.

   For each expression, choose a sub-rectangle whose area, in square units, matches the expression.

   a. 3 · (0.6)

   b. (0.4) · 2

   c. (0.4) · (0.6)

   d. 3 · 2

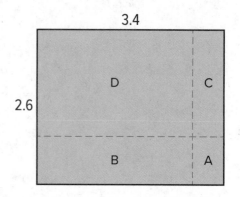

2. Here is an area diagram that represents (3.1) · (1.4).

   a. Find the areas of sub-rectangles A and B.

   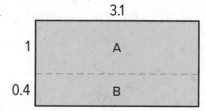

   b. What is the area of the 3.1 by 1.4 rectangle?

NAME _____ DATE _____ PERIOD _____

3. Draw an area diagram to find (0.36) • (0.53). Label and organize
your work so that it can be followed by others.

4. Find each product. Show your reasoning.
   a. (2.5) • (1.4)

   b. (0.64) • (0.81)

5. Complete the calculations so that each shows the correct sum. (Lesson 5-3)

a.   2 . 3 ☐
      + ☐ . 6 4
      ―――――
        9 . ☐ 5

b.   2 . 3 ☐
      + ☐ . 6 4
      ―――――
        9 . ☐ 2

c.   4 . 3 ☐
      + ☐ . 1 5
      ―――――
        6 . ☐ 2

d.   1 . 5 ☐
      + ☐ . 3 8
      ―――――
        1 . ☐ 4

6. Diego bought 12 mini muffins for $4.20. (Lesson 2-12)

Number of Mini Muffins	Price in Dollars
12	4.20

   a. At this rate, how much would Diego pay for 4 mini muffins?

   b. How many mini muffins could Diego buy with $3.00? Explain or show your reasoning. If you get stuck, consider using the table.

Lesson 5-8

Calculating Products of Decimals

NAME _____ DATE _____ PERIOD _____

Learning Goal Let's multiply decimals.

 ## Warm Up
8.1 Number Talk: Twenty Times a Number

Evaluate mentally.

1. $20 \cdot 5$

2. $20 \cdot (0.8)$

3. $20 \cdot (0.04)$

4. $20 \cdot (5.84)$

 ## Activity
8.2 Calculating Products of Decimals

1. A common way to find a product of decimals is to calculate a product of whole numbers, then place the decimal point in the product.

   Here is an example for $(2.5) \cdot (1.2)$.

   Use what you know about decimals and place value to explain why the decimal point of the product is placed where it is.

$$\begin{array}{r} 2\ 5 \\ \times \quad 1\ 2 \\ \hline 5\ 0 \\ +\quad 2\ 5\ 0 \\ \hline 3\ 0\ 0 \end{array}$$

$$25 \cdot 12 = 300$$

$$(2.5) \cdot (1.2) = 3.00$$

2. Use the method shown in the first question to calculate each product.

   a. $(4.6) \cdot (0.9)$

   b. $(16.5) \cdot (0.7)$

3. Use area diagrams to check your earlier calculations. For each problem:

   - Decompose each number into its base-ten units and write them in the boxes on each side of the rectangle.

   - Write the area of each lettered region in the diagram. Then find the area of the entire rectangle. Show your reasoning.

   a. $(4.6) \cdot (0.9)$

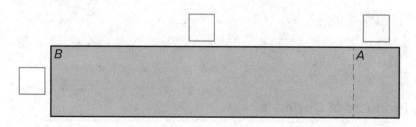

   b. $(16.5) \cdot (0.7)$

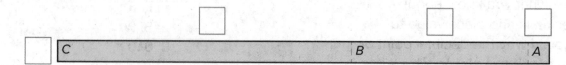

4. About how many centimeters are in 6.25 inches if 1 inch is about 2.5 centimeters? Show your reasoning.

NAME _____ DATE _____ PERIOD _____

Activity

8.3 Practicing Multiplication of Decimals

1. Calculate each product. Show your reasoning. If you get stuck, consider drawing an area diagram to help.

   a. $(5.6) \cdot (1.8)$

   b. $(0.008) \cdot (7.2)$

2. A rectangular playground is 18.2 meters by 12.75 meters.

   a. Find its area in square meters. Show your reasoning.

   b. If 1 meter is approximately 3.28 feet, what are the approximate side lengths of the playground in feet? Show your reasoning.

Are you ready for more?

1. Write the following expressions as decimals.

   a. $1 - 0.1$

   b. $1 - 0.1 + 10 - 0.01$

   c. $1 - 0.1 + 10 - 0.01 + 100 - 0.001$

2. Describe the decimal that results as this process continues.

3. What would happen to the decimal if all of the positive and negative signs became multiplication symbols? Explain your reasoning.

Summary
Calculating Products of Decimals

We can use 84 · 43 and what we know about place value to find (8.4) · (4.3).

Since 8.4 is 84 tenths and 4.3 is 43 tenths, then:

$$(8.4) \cdot (4.3) = \frac{84}{10} \cdot \frac{43}{10}$$

$$= \frac{84 \cdot 43}{100}$$

That means we can compute 84 · 43 and then divide by 100 to find (8.4) · (4.3).

$$84 \cdot 43 = 3612$$

$$(8.4) \cdot (4.3) = 36.12$$

Using fractions such as $\frac{1}{10}$, $\frac{1}{100}$, and $\frac{1}{1,000}$ allows us to find the product of two decimals using the following steps:

- Write each decimal factor as a product of a whole number and a fraction.
- Multiply the whole numbers.
- Multiply the fractions.
- Multiply the products of the whole numbers and fractions.

We know multiplying by fractions such as $\frac{1}{10}$, $\frac{1}{100}$, and $\frac{1}{1,000}$ is the same as dividing by 10, 100, and 1,000, respectively. This means we can move the decimal point in the whole-number product to the left the appropriate number of spaces to correctly place the decimal point.

NAME _____ DATE _____ PERIOD _____

Practice
Calculating Products of Decimals

1. Here are an unfinished calculation of (0.54) • (3.8) and a 0.54-by-3.8 rectangle.

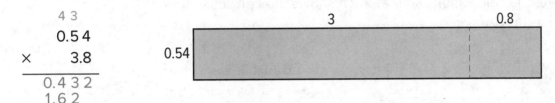

```
    4 3
  0.5 4
×   3.8
─────────
  0.4 3 2
  1.6 2
```

a. Which part of the rectangle has an area of 0.432? Which part of the rectangle has an area of 1.62? Show your reasoning.

b. What is (0.54) • (3.8)?

2. Explain how the product of 3 and 65 could be used to find (0.03) • (0.65).

3. Use vertical calculation to find each product.

   a. (5.4) • (2.4)   b. (1.67) • (3.5)

4. A pound of blueberries costs $3.98 and a pound of clementines costs $2.49. What is the combined cost of 0.6 pound of blueberries and 1.8 pounds of clementines? Round your answer to the nearest cent.

5. Complete the calculations so that each shows the correct sum or difference. (Lesson 5-3)

a.
```
  2 . 1 4 □
+ 1 . □ 2 5
─────────
  □ . 8 6 5
```

b.
```
  2 9 . □□
+    1 . 4 2
─────────
  □□ . 4 1
```

c.
```
  6 . 1 □□
− 1 . 0 0 9
─────────
  □ . □ 4 8
```

6. Which has a greater value: 7.4 − 0.0022 or 7.39 − 0.0012? Show your reasoning. (Lesson 5-4)

7. Andre is planting saplings (baby trees). It takes him 30 minutes to plant 3 saplings. If each sapling takes the same amount of time to plant, how long will it take Andre to plant 14 saplings? If you get stuck, consider using the table. (Lesson 2-12)

Number of Saplings	Time in Minutes
3	30
1	
14	

Lesson 5-9

Using the Partial Quotients Method

NAME _____ DATE _____ PERIOD _____

Learning Goal Let's divide whole numbers.

Warm Up
9.1 Using Base-Ten Diagrams to Calculate Quotients

Elena used base-ten diagrams to find $372 \div 3$. She started by representing 372.

3 hundreds   7 tens   2 ones

She made 3 groups, each with 1 hundred. Then, she put the tens and ones in each of the 3 groups. Here is her diagram for $372 \div 3$.

hundreds   tens   ones

Discuss with a partner:

1. Elena's diagram for 372 has 7 tens. The one for $372 \div 3$ has only 6 tens. Why?

2. Where did the extra ones (small squares) come from?

Activity

9.2 Using the Partial Quotients Method to Calculate Quotients

1. Andre calculated 657 ÷ 3 using a method that was different from Elena's.

He started by writing the dividend (657) and the divisor (3).	He then subtracted 3 groups of different amounts from 657, starting with 3 groups of 200 . . .	. . . then 3 groups of 10, and then 3 groups of 9.	Andre calculated 200 + 10 + 9 and then wrote 219.
$3\overline{)6\ 5\ 7}$	$\begin{array}{r} 2\ 0\ 0 \\ 3\overline{)6\ 5\ 7} \\ -6\ 0\ 0 \\ \hline 5\ 7 \end{array}$	$\begin{array}{r} 9 \\ 1\ 0 \\ 2\ 0\ 0 \\ 3\overline{)6\ 5\ 7} \\ -6\ 0\ 0 \\ \hline 5\ 7 \\ -3\ 0 \\ \hline 2\ 7 \\ -2\ 7 \\ \hline 0 \end{array}$	$\begin{array}{r} \boxed{2\ 1\ 9} \\ 9 \\ 1\ 0 \\ 2\ 0\ 0 \\ 3\overline{)6\ 5\ 7} \\ -6\ 0\ 0 \\ \hline 5\ 7 \\ -3\ 0 \\ \hline 2\ 7 \\ -2\ 7 \\ \hline 0 \end{array}$

a. Andre subtracted 600 from 657. What does the 600 represent?

b. Andre wrote 10 above the 200, and then subtracted 30 from 57. How is the 30 related to the 10?

c. What do the numbers 200, 10, and 9 represent?

d. What is the meaning of the 0 at the bottom of Andre's work?

2. How might Andre calculate 896 ÷ 4? Explain or show your reasoning.

NAME _____ DATE _____ PERIOD _____

Activity
9.3 What's the Quotient?

1. Find the quotient of 1,332 ÷ 9 using one of the methods you have seen so far. Show your reasoning.

2. Find each quotient and show your reasoning. Use the partial quotients method at least once.

   a. 1,115 ÷ 5

   b. 665 ÷ 7

   c. 432 ÷ 16

We can find the quotient $345 \div 3$ in different ways.

One way is to use a base-ten diagram to represent the hundreds, tens, and ones and to create equal-sized groups.

hundreds   tens   ones

We can think of the division by 3 as splitting up 345 into 3 equal groups.

hundreds   tens   ones

group 1

group 2

group 3

Each group has 1 hundred, 1 ten, and 5 ones, so $345 \div 3 = 115$. Notice that in order to split 345 into 3 equal groups, one of the tens had to be unbundled or decomposed into 10 ones.

NAME _____   DATE _____   PERIOD _____

Another way to divide 345 by 3 is by using the partial quotients method, in which we keep subtracting 3 groups of some amount from 345.

$$
\begin{array}{r}
\boxed{1\ 1\ 5} \\
5 \\
1\ 0 \\
1\ 0\ 0 \\
3\overline{)3\ 4\ 5} \\
-3\ 0\ 0 \quad \longleftarrow \text{3 groups of 100} \\
\hline
4\ 5 \\
-3\ 0 \quad \longleftarrow \text{3 groups of 10} \\
\hline
1\ 5 \\
-1\ 5 \quad \longleftarrow \text{3 groups of 5} \\
\hline
0
\end{array}
$$

$$
\begin{array}{r}
\boxed{1\ 1\ 5} \\
5\ 0 \\
5\ 0 \\
1\ 5 \\
3\overline{)3\ 4\ 5} \\
-\ 4\ 5 \quad \longleftarrow \text{3 groups of 15} \\
\hline
3\ 0\ 0 \\
-1\ 5\ 0 \quad \longleftarrow \text{3 groups of 50} \\
\hline
1\ 5\ 0 \\
-1\ 5\ 0 \quad \longleftarrow \text{3 groups of 50} \\
\hline
0
\end{array}
$$

In the calculation on the left, first we subtract 3 groups of 100, then 3 groups of 10, and then 3 groups of 5. Adding up the partial quotients (100 + 10 + 5) gives us 115.

The calculation on the right shows a different amount per group subtracted each time (3 groups of 15, 3 groups of 50, and 3 more groups of 50), but the total amount in each of the 3 groups is still 115. There are other ways of calculating 345 ÷ 3 using the partial quotients method.

Both the base-ten diagrams and partial quotients methods are effective. If, however, the dividend and divisor are large, as in 1,248 ÷ 26, then the base-ten diagrams will be time-consuming.

1. Here is one way to find 2,105 ÷ 5 using partial quotients.

$$
\begin{array}{r}
\boxed{4\ 2\ 1} \\
1 \\
2\ 0 \\
4\ 0\ 0 \\
5\overline{)2\ 1\ 0\ 5} \\
-\ 2\ 0\ 0\ 0 \\
\hline
1\ 0\ 5 \\
-\quad 1\ 0\ 0 \\
\hline
5 \\
-\quad\ 5 \\
\hline
0
\end{array}
$$

Show a different way of using partial quotients to divide 2,105 by 5.

NAME _____ DATE _____ PERIOD _____

2. Andre and Jada both found 657 ÷ 3 using the partial quotients method, but they did the calculations differently, as shown here.

Andre's Work	**Jada's Work**

```
        2 1 9                    2 1 9
            9                        9
          1 0                      6 0
        2 0 0                    1 0 0
      ┌─────                       5 0
     3)6 5 7                   ┌─────
      - 6 0 0                 3)6 5 7
        ─────                  - 1 5 0
          5 7                    ─────
        - 3 0                    5 0 0
        ─────                  - 3 0 0
          2 7                    ─────
        - 2 7                    2 0 7
        ─────                  - 1 8 0
            0                    ─────
                                  2 7
                                - 2 7
                                ─────
                                    0
```

a. How is Jada's work the same as Andre's work? How is it different?

b. Explain why they have the same answer.

3. Which might be a better way to evaluate 1,150 ÷ 46: drawing base-ten diagrams or using the partial quotients method? Explain your reasoning.

4. Here is an incomplete calculation of 534 ÷ 6.

   Write the missing numbers (marked with "?")
   that would make the calculation complete.

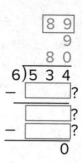

5. Use the partial quotients method to find 1,032 ÷ 43.

6. Which of the polygons has the greatest area? (Lesson 5-8)

   (A.) A rectangle that is 3.25 inches wide and 6.1 inches long.

   (B.) A square with side length of 4.6 inches.

   (C.) A parallelogram with a base of 5.875 inches and a height of 3.5 inches.

   (D.) A triangle with a base of 7.18 inches and a height of 5.4 inches.

7. One micrometer is a millionth of a meter. A certain spider web is
   4 micrometers thick. A fiber in a shirt is 1 hundred-thousandth of
   a meter thick. (Lesson 5-4)

   a. Which is wider, the spider web or the fiber? Explain your reasoning.

   b. How many meters wider?

Lesson 5-10

Using Long Division

NAME _____ DATE _____ PERIOD _____

Learning Goal Let's use long division.

 Warm Up
10.1 Number Talk: Estimating Quotients

Estimate these quotients mentally.

1. $500 \div 7$

2. $1{,}394 \div 9$

Activity

10.2 Lin Uses Long Division

Lin has a method of calculating quotients that is different from Elena's method and Andre's method. Here is how she found the quotient of 657 ÷ 3:

Lin arranged the numbers for vertical calculations. Her plan was to divide each digit of 657 into 3 groups, starting with the 6 hundreds.	There are 3 groups of 2 in 6, so Lin wrote 2 at the top and subtracted 6 from the 6, leaving 0. Then, she brought down the 5 tens of 657.	There are 3 groups of 1 in 5, so she wrote 1 at the top and subtracted 3 from 5, which left a remainder of 2.	She brought down the 7 ones of 657 and wrote it next to the 2, which made 27. There are 3 groups of 9 in 27, so she wrote 9 at the top and subtracted 27, leaving 0.
| ```
  _____
3)6 5 7
``` | ```
      2
  _____
3)6 5 7
 −6 ↓
  _____
  0 5
``` | ```
    2 1
  _____
3)6 5 7
 −6
  ____
   5
 − 3
  ____
   2
``` | ```
    2 1 9
  _____
3)6 5 7
 −6 |
  ____
   5
 − 3 ↓
  ____
   2 7
 − 2 7
  ____
     0
``` |

1. Discuss with your partner how Lin's method is similar to and different from drawing base-ten diagrams or using the partial quotients method.

   a. Lin subtracted 3 · 2, then 3 · 1, and lastly 3 · 9. Earlier, Andre subtracted 3 · 200, then 3 · 10, and lastly 3 · 9. Why did they have the same quotient?

   b. In the third step, why do you think Lin wrote the 7 next to the remainder of 2 rather than adding 7 and 2 to get 9?

2. Lin's method is called **long division**. Use this method to find the following quotients. Check your answer by multiplying it by the divisor.

   a. 846 ÷ 3   b. 1,816 ÷ 4   c. 768 ÷ 12

NAME _____ DATE _____ PERIOD _____

Activity
10.3 Dividing Whole Numbers

1. Find each quotient.

   a. $633 \div 3$

   b. $1001 \div 7$

   c. $2996 \div 14$

2. Here is Priya's calculation of $906 \div 3$.

   ```
         3 2 0
      ┌─────────
   3 )  9 0 6
    −  6
      ─────────
         0 6
    −       6
      ─────────
           0
   ```

   a. Priya wrote 320 for the value of $906 \div 3$. Check her answer by multiplying it by 3. What product do you get and what does it tell you about Priya's answer?

   b. Describe Priya's mistake, then show the correct calculation and answer.

Long division is another method for calculating quotients. It relies on place value to perform and record the division.

When we use long division, we work from left to right and with one digit at a time, starting with the leftmost digit of the dividend. We remove the largest group possible each time, using the placement of the digit to indicate the size of each group. Here is an example of how to find 948 ÷ 3 using long division.

```
      3 1 6
   3)9 4 8
    - 9           ◄——— 3 groups of 3 (hundreds)
      4
    -   3         ◄——— 3 groups of 1 (ten)
      1 8
    - 1 8         ◄——— 3 groups of 6 (ones)
        0
```

- We start by dividing 9 hundreds into 3 groups, which means 3 hundreds in each group. Instead of writing 300, we simply write 3 in the hundreds place, knowing that it means 3 hundreds.

- There are no remaining hundreds, so we work with the tens. We can make 3 groups of 1 ten in 4 tens, so we write 1 in the tens place above the 4 of 948. Subtracting 3 tens from 4 tens, we have a remainder of 1 ten.

- We know that 1 ten is 10 ones. Combining these with the 8 ones from 948, we have 18 ones. We can make 3 groups of 6, so we write 6 in the ones place.

In total, there are 3 groups of 3 hundreds, 1 ten, and 6 ones in 948, so 948 ÷ 3 = 316.

Glossary
_____

long division

NAME _____ DATE _____ PERIOD _____

Practice
Using Long Division

1. Kiran is using long division to find 696 ÷ 12.

   He starts by dividing 69 by 12. In which decimal place should Kiran place the first digit of the quotient (5)?

   $$12\overline{)696}$$

   (A.) hundreds

   (B.) tens

   (C.) ones

   (D.) tenths

2. Here is a long-division calculation of 917 ÷ 7.

   a. There is a 7 under the 9 of 917. What does this 7 represent?

   b. What does the subtraction of 7 from 9 mean?

   c. Why is a 1 written next to the 2 from 9 − 7?

   $$
   \begin{array}{r}
   1\ 3\ 1 \\
   7\overline{)9\ 1\ 7} \\
   -\ 7 \\
   \hline
   2\ 1 \\
   -\ 2\ 1 \\
   \hline
   7 \\
   -\ 7 \\
   \hline
   0
   \end{array}
   $$

3. Han's calculation of $972 \div 9$ is shown here.

  a. Find $180 \cdot 9$.

```
      1 8 0
  9)9 7 2
   - 9
    ‾‾‾‾
      7 2
    - 7 2
     ‾‾‾‾
        0
      - 0
     ‾‾‾‾
        0
```

  b. Use your calculation of $180 \cdot 9$ to explain how you know Han has made a mistake.

  c. Identify and correct Han's mistake.

4. Find each quotient.

  a. $5)\overline{4\,6\,5}$

  b. $12)\overline{9\,2\,4}$

  c. $3)\overline{1\,1\,0\,7}$

5. One ounce of a yogurt contains of 1.2 grams of sugar. How many grams of sugar are in 14.25 ounces of yogurt? **(Lesson 5-7)**

  (A.) 0.171 grams

  (C.) 17.1 grams

  (B.) 1.71 grams

  (D.) 171 grams

6. The mass of one coin is 16.718 grams. The mass of a second coin is 27.22 grams. How much greater is the mass of the second coin than the first? Show your reasoning. **(Lesson 5-4)**

Lesson 5-11

Dividing Numbers That Result in Decimals

NAME _____ DATE _____ PERIOD _____

Learning Goal Let's find quotients that are not whole numbers.

 Warm Up
11.1 Number Talk: Evaluating Quotients

Find the quotients mentally.

1. 400 ÷ 8

2. 80 ÷ 8

3. 16 ÷ 8

4. 496 ÷ 8

Activity

11.2 Keep Dividing

Mai used base-ten diagrams to calculate 62 ÷ 5. She started by representing 62.

6 tens   2 ones

She then made 5 groups, each with 1 ten. There was 1 ten left. She unbundled it into 10 ones and distributed the ones across the 5 groups.

tens   ones   tenths

Here is Mai's diagram for 62 ÷ 5.

1. Discuss these questions with a partner and write down your answers:

   a. Mai should have a total of 12 ones, but her diagram shows only 10. Why?

   b. She did not originally have tenths, but in her diagram each group has 4 tenths. Why?

   c. What value has Mai found for 62 ÷ 5? Explain your reasoning.

2. Find the quotient of 511 ÷ 5 by drawing base-ten diagrams or by using the partial quotients method. Show your reasoning. If you get stuck, work with your partner to find a solution.

3. Four students share a $271 prize from a science competition. How much does each student get if the prize is shared equally? Show your reasoning.

NAME _____ DATE _____ PERIOD _____

Activity

11.3 Using Long Division to Calculate Quotients

1. Here is how Lin calculated 62 ÷ 5.

| Lin set up the numbers for long division. | She subtracted 5 times 1 from the 6, which leaves a remainder of 1. She wrote the 2 from 62 next to the 1, which made 12, and subtracted 5 times 2 from 12. | Lin drew a vertical line and a decimal point, separating the ones and tenths place. 12 − 10 is 2. She wrote 0 to the right of the 2, which made 20. | Lastly, she subtracted 5 times 4 from 20, which left no remainder. At the top, she wrote 4 next to the decimal point. |

$$
\begin{array}{r}
5)\overline{6\ 2}
\end{array}
$$

$$
\begin{array}{r}
1\\
5)\overline{6\ 2}\\
-5\\
\hline
1\ 2\\
-1\ 0\\
\hline
2
\end{array}
$$

$$
\begin{array}{r}
1\ 2.\\
5)\overline{6\ 2}\\
-5\\
\hline
1\ 2\\
-1\ 0\\
\hline
2\,0
\end{array}
$$

$$
\begin{array}{r}
1\ 2.4\\
5)\overline{6\ 2}\\
-5\\
\hline
1\ 2\\
-1\ 0\\
\hline
2\,0\\
-2\,0\\
\hline
0
\end{array}
$$

Discuss with your partner:

a. Lin put a 0 after the remainder of 2. Why? Why does this 0 not change the value of the quotient?

b. Lin subtracted 5 groups of 4 from 20. What value does the 4 in the quotient represent?

c. What value did Lin find for 62 ÷ 5?

2. Use long division to find the value of each expression. Then pause so your teacher can review your work.

a. 126 ÷ 8

b. 90 ÷ 12

3. Use long division to show that:

   a. $5 \div 4$, or $\frac{5}{4}$, is 1.25.

   b. $4 \div 5$, or $\frac{4}{5}$, is 0.8.

   c. $1 \div 8$, or $\frac{1}{8}$, is 0.125.

   d. $1 \div 25$, or $\frac{1}{25}$, is 0.04.

4. Noah said we cannot use long division to calculate $10 \div 3$ because there will always be a remainder.

   a. What do you think Noah meant by "there will always be a remainder"?

   b. Do you agree with him? Explain your reasoning.

NAME _____ DATE _____ PERIOD _____

Summary
Dividing Numbers that Result in Decimals

Dividing a whole number by another whole number does not always produce a whole-number quotient. Let's look at 86 ÷ 4, which we can think of as dividing 86 into 4 equal groups.

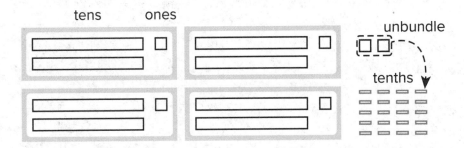

We can see in the base-ten diagram that there are 4 groups of 21 in 86 with 2 ones left over. To find the quotient, we need to distribute the 2 ones into the 4 groups. To do this, we can unbundle or decompose the 2 ones into 20 tenths, which enables us to put 5 tenths in each group.

Once the 20 tenths are distributed, each group will have 2 tens, 1 one, and 5 tenths, so 86 ÷ 4 = 21.5.

We can also calculate 86 ÷ 4 using long division.

$$
\begin{array}{r}
2\ 1.5 \\
4\overline{)8\ 6} \\
-8 \\
\hline
6 \\
-4 \\
\hline
2\,0 \\
-2\,0 \\
\hline
0
\end{array}
$$

The calculation shows that, after removing 4 groups of 21, there are 2 ones remaining. We can continue dividing by writing a 0 to the right of the 2 and thinking of that remainder as 20 tenths, which can then be divided into 4 groups.

To show that the quotient we are working with now is in the tenth place, we put a decimal point to the right of the 1 (which is in the ones place) at the top. It may also be helpful to draw a vertical line to separate the ones and the tenths.

There are 4 groups of 5 tenths in 20 tenths, so we write 5 in the tenths place at the top. The calculation likewise shows 86 ÷ 4 = 21.5.

Practice

Dividing Numbers that Result in Decimals

1. Use long division to show that the fraction and decimal in each pair are equal.

   a. $\frac{3}{4}$ and 0.75

   b. $\frac{3}{50}$ and 0.06

   c. $\frac{7}{25}$ and 0.28

2. Mai walked $\frac{1}{8}$ of a 30-mile walking trail. How many miles did Mai walk? Explain or show your reasoning.

NAME _____ DATE _____ PERIOD _____

3. Use long division to find each quotient. Write your answer as a decimal.

   a. $99 \div 12$ **b.** $216 \div 5$ **c.** $1{,}988 \div 8$

4. Tyler reasoned: "$\frac{9}{25}$ is equivalent to $\frac{18}{50}$ and to $\frac{36}{100}$, so the decimal of $\frac{9}{25}$ is 0.36."

   a. Use long division to show that Tyler is correct.

   b. Is the decimal of $\frac{18}{50}$ also 0.36? Use long division to support your answer.

5. Complete the calculations so that each shows the correct difference.

(Lesson 5-4)

a.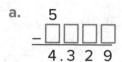

b.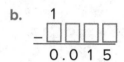

c.

$$
\begin{array}{r}
1 \\
-\ \square\square\square\square \\
\hline
0.8\ 6\ 3
\end{array}
$$

6. Use the equation $124 \cdot 15 = 1{,}860$ and what you know about fractions, decimals, and place value to explain how to place the decimal point when you compute $(1.24) \cdot (0.15)$. (Lesson 5-6)

Lesson 5-12

Dividing Decimals by Whole Numbers

NAME _____ DATE _____ PERIOD _____

Learning Goal Let's divide decimals by whole numbers.

 ## Warm Up
12.1 Number Talk: Dividing by 4

Find each quotient mentally.

1. 80 ÷ 4

2. 12 ÷ 4

3. 1.2 ÷ 4

4. 81.2 ÷ 4

Activity

12.2 Using Diagrams to Represent Division

To find 53.8 ÷ 4 using diagrams, Elena began by representing 53.8.

5 tens   3 ones   8 tenths

She placed 1 ten into each group, unbundled the remaining 1 ten into 10 ones, and went on distributing the units.

This diagram shows Elena's initial placement of the units and the unbundling of 1 ten.

tens   ones   tenths   hundredths

group 1

group 2

group 3

group 4

1. Complete the diagram by continuing the division process. How would you use the available units to make 4 equal groups?

   As the units get placed into groups, show them accordingly and cross out those pieces from the bottom. If you unbundle a unit, draw the resulting pieces.

   unbundle

2. What value did you find for 53.8 ÷ 4? Be prepared to explain your reasoning.

3. Use long division to find 53.8 ÷ 4. Check your answer by multiplying it by the divisor 4.

4. Use long division to find 77.4 ÷ 5. If you get stuck, you can draw diagrams or use another method.

NAME _____ DATE _____ PERIOD _____

Are you ready for more?

A distant, magical land uses jewels for their bartering system. The jewels are valued and ranked in order of their rarity. Each jewel is worth 3 times the jewel immediately below it in the ranking. The ranking is red, orange, yellow, green, blue, indigo, and violet. So a red jewel is worth 3 orange jewels, a green jewel is worth 3 blue jewels, and so on.

A group of 4 craftsmen are paid 1 of each jewel. If they split the jewels evenly amongst themselves, which jewels does each craftsman get?

Activity

12.3 Dividends and Divisors

Analyze the dividends, divisors, and quotients in the calculations, and then answer the questions.

$$
\begin{array}{r} 24 \\ 3\overline{)72} \\ -6\downarrow \\ \hline 12 \\ -12 \\ \hline 0 \end{array}
\qquad
\begin{array}{r} 24 \\ 30\overline{)720} \\ -60\downarrow \\ \hline 120 \\ -120 \\ \hline 0 \end{array}
\qquad
\begin{array}{r} 24 \\ 300\overline{)7200} \\ -600\downarrow \\ \hline 1200 \\ 1200 \\ \hline 0 \end{array}
\qquad
\begin{array}{r} 24 \\ 3000\overline{)72000} \\ -6000\downarrow \\ \hline 12000 \\ -12000 \\ \hline 0 \end{array}
$$

1. Complete each sentence. In the calculations shown:

   a. Each dividend is _____ times the dividend to the left of it.

   b. Each divisor is _____ times the divisor to the left of it.

   c. Each quotient is _____ the quotient to the left of it.

2. Suppose we are writing a calculation to the right of 72,000 ÷ 3,000. Which expression has a quotient of 24? Be prepared to explain your reasoning.

   (A.) 72,000 ÷ 30,000

   (B.) 720,000 ÷ 300,000

   (C.) 720,000 ÷ 30,000

   (D.) 720,000 ÷ 3,000

3. Suppose we are writing a calculation to the left of 72 ÷ 3. Write an expression that would also give a quotient of 24. Be prepared to explain your reasoning.

4. Decide which of the following expressions would have the same value as 250 ÷ 10. Be prepared to share your reasoning.

(A.) 250 ÷ 0.1   (D.) 2.5 ÷ 0.1

(B.) 25 ÷ 1   (E.) 2,500 ÷ 100

(C.) 2.5 ÷ 1   (F.) 0.25 ÷ 0.01

Summary
Dividing Decimals by Whole Numbers

We know that fractions such as $\frac{6}{4}$ and $\frac{60}{40}$ are equivalent because:

- The numerator and denominator of $\frac{60}{40}$ are each 10 times those of $\frac{6}{4}$.

- Both fractions can be simplified to $\frac{3}{2}$.

- 600 divided by 400 is 1.5, and 60 divided by 40 is also 1.5.

Just like fractions, division expressions can be equivalent.

For example, the expressions 540 ÷ 90 and 5,400 ÷ 900 are both equivalent to 54 ÷ 9 because:

- They all have a quotient of 6.

- The dividend and the divisor in 540 ÷ 90 are each 10 times the dividend and divisor in 54 ÷ 9. Those in 5,400 ÷ 900 are each 100 times the dividend and divisor in 54 ÷ 9. In both cases, the quotient does not change.

This means that an expression such as 5.4 ÷ 0.9 also has the same value as 54 ÷ 9. Both the dividend and divisor of 5.4 ÷ 0.9 are $\frac{1}{10}$ of those in 54 ÷ 9.

In general, multiplying a dividend and a divisor by the same number does not change the quotient. Multiplying by powers of 10 (e.g., 10, 100, 1,000, etc.) can be particularly useful for dividing decimals, as we will see in an upcoming lesson.

NAME _____ DATE _____ PERIOD _____

Practice
Dividing Decimals by Whole Numbers

1. Here is a diagram representing a base-ten number. The large
rectangle represents a unit that is 10 times the value of the square. The
square represents a unit that is 10 times the value of the small rectangle.

⬜▭ ▫▫▫ ▬ ▬

Here is a diagram showing the number being divided into
5 equal groups.

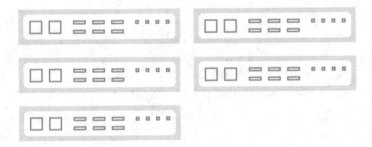

a. If a large rectangle represents 1,000, what division problem
did the second diagram show? What is its answer?

b. If a large rectangle represents 100, what division problem did the
second diagram show? What is its answer?

c. If a large rectangle represents 10, what division problem did the
second diagram show? What is its answer?

2. Respond to each of the following.

a. Explain why all of these expressions have the same value.

$4.5 \div 0.09$ $\qquad$ $45 \div 0.9$ $\qquad$ $450 \div 9$ $\qquad$ $4{,}500 \div 90$

b. What is the common value?

3. Use long division to find each quotient.

    a. $7.89 \div 2$ b. $39.54 \div 3$ c. $0.176 \div 5$

4. Four students set up a lemonade stand. At the end of the day, their profit is $17.52. How much money do they each have when the profit is split equally? Show or explain your reasoning.

5. Respond to each question. **(Lesson 5-8)**

    a. A standard sheet of paper in the United States is 11 inches long and 8.5 inches wide. Each inch is 2.54 centimeters. How long and wide is a standard sheet of paper in centimeters?

    b. A standard sheet of paper in Europe is 21.0 cm wide and 29.7 cm long. Which has the greater area, the standard sheet of paper in the United States or the standard sheet of paper in Europe? Explain your reasoning.

Lesson 5-13

Dividing Decimals by Decimals

NAME _____ DATE _____ PERIOD _____

Learning Goal Let's divide decimals by decimals.

 ## Warm Up
13.1 Same Values

1. Use long division to find the value of $5.04 \div 7$.

2. Select all of the quotients that have the same value as $5.04 \div 7$. Be prepared to explain how you know.

   A. $5.04 \div 70$ C. $504,000 \div 700$

   B. $50.4 \div 70$ D. $504,000 \div 700,000$

 ## Activity
13.2 Placing Decimal Points in Quotients

1. Think of one or more ways to find $3 \div 0.12$. Show your reasoning.

2. Find $1.8 \div 0.004$. Show your reasoning. If you get stuck, think about what equivalent division expression you could write.

3. Diego said, "To divide decimals, we can start by moving the decimal point in both the dividend and divisor by the same number of places and in the same direction. Then we find the quotient of the resulting numbers."

   Do you agree with Diego? Use the division expression $7.5 \div 1.25$ to support your answer.

Can we create an equivalent division expression by multiplying both the dividend and divisor by a number that is *not* a multiple of 10 (for example: 4, 20, or $\frac{1}{2}$)? Would doing so produce the same quotient? Explain or show your reasoning.

 ## Activity

13.3 Two Ways to Calculate Quotients of Decimals

1. Here are two calculations of 48.78 ÷ 9. Work with your partner to answer the following questions.

Calculation A	Calculation B
| ```
      5 4 2
   9)4 8.7 8
    -4 5
      3 7
     -3 6
        1 8
       -1 8
          0
``` | ```
         5 4 2
  900)4 8 7 8
     -4 5 0 0
        3 7 8 0
       -3 6 0 0
          1 8 0 0
         -1 8 0 0
              0
``` |

a. How are the two calculations the same? How are they different?

b. Look at Calculation A. Explain how you can tell that the 36 means "36 tenths" and the 18 means "18 hundredths."

c. Look at Calculation B. What do the 3600 and 1800 mean?

d. We can think of 48.78 ÷ 9 = 5.42 as saying, "There are 9 groups of 5.42 in 48.78." We can think of 4,878 ÷ 900 = 5.42 as saying, "There are 900 groups of 5.42 in 4,878." How might we show that both statements are true?

NAME _____ DATE _____ PERIOD _____

2. a. Explain why 51.2 ÷ 6.4 has the same value as 5.12 ÷ 0.64.

   b. Write a division expression that has the same value as 51.2 ÷ 6.4 but is easier to use to find the value. Then, find the value using long division.

 ## Activity

13.4 Practicing Division with Decimals

Find each quotient. Discuss your quotients with your group and agree on the correct answers. Consult your teacher if the group can't agree.

1. 106.5 ÷ 3

2. 58.8 ÷ 0.7

3. 257.4 ÷ 1.1

4. Mai is making friendship bracelets. Each bracelet is made from 24.3 cm of string. If she has 170.1 cm of string, how many bracelets can she make? Explain or show your reasoning.

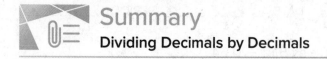
One way to find a quotient of two decimals is to multiply each decimal by a power of 10 so that both products are whole numbers.

If we multiply both decimals by the same power of 10, this does not change the value of the quotient.

For example, the quotient $7.65 \div 1.2$ can be found by multiplying the two decimals by 10 (or by 100) and instead finding $76.5 \div 12$ or $765 \div 120$.

To calculate $765 \div 120$, which is equivalent to $76.5 \div 12$, we could use base-ten diagrams, partial quotients, or long division.

Here is the calculation with long division.

```
           6 3 7 5
     120)7 6 5
       - 7 2 0
           4 5 0
         - 3 6 0
             9 0 0
           - 8 4 0
               6 0 0
             - 6 0 0
                   0
```

NAME _____ DATE _____ PERIOD _____

 ## Practice
Dividing Decimals by Decimals

1. A student said, "To find the value of 109.2 ÷ 6, I can divide 1,092 by 60."

   a. Do you agree with her? Explain your reasoning.

   b. Calculate the quotient of 109.2 ÷ 6 using any method of your choice.

2. Here is how Han found 31.59 ÷ 13:

   a. At the second step, Han subtracts 52 from 55. How do you know that these numbers represent tenths?

```
         2 ¦ 4 3
  13)3 1 ¦ 5 9
   − 2 6 ¦
   ──────
       5 ¦ 5
     − 5 ¦ 2
     ──────
         ¦ 3 9
       − ¦ 3 9
       ──────
         ¦   0
```

   b. At the third step, Han subtracts 39 from 39. How do you know that these numbers represent hundredths?

   c. Check that Han's answer is correct by calculating the product of 2.43 and 13.

3. Respond to each of the following.

    a. Write two division expressions that have the same value as $61.12 \div 3.2$.

    b. Find the value of $61.12 \div 3.2$. Show your reasoning.

4. A bag of pennies weighs 5.1 kilograms. Each penny weighs 2.5 grams. About how many pennies are in the bag?

    (A.) 20

    (B.) 200

    (C.) 2,000

    (D.) 20,000

NAME _____ DATE _____ PERIOD _____

5. Find each difference. If you get stuck, consider drawing a diagram.
(Lesson 5-3)

a. $2.5 - 1.6$

b. $0.72 - 0.4$

c. $11.3 - 1.75$

d. $73 - 1.3$

6. Plant B is $6\frac{2}{3}$ inches tall. Plant C is $4\frac{4}{15}$ inches tall. Complete the sentences and show your reasoning. (Lesson 4-12)

   a. Plant C is _____ times as tall as Plant B.

   b. Plant C is _____ inches _____ (taller or shorter) than Plant B.

7. At a school, 460 of the students walk to school. (Lesson 3-15)

   a. The number of students who take public transit is 20% of the number of students who walk. How many students take public transit?

   b. The number of students who bike to school is 5% of the number of students who walk. How many students bike to school?

   c. The number of students who ride the school bus is 110% of the number of students who walk. How many students ride the school bus?

Lesson 5-14

Using Operations on Decimals to Solve Problems

NAME _____ DATE _____ PERIOD _____

Learning Goal Let's solve some problems using decimals.

 Warm Up

14.1 Close Estimates

For each expression, choose the best estimate of its value.

1. 76.2 ÷ 15

   • 0.5

   • 5

   • 50

2. 56.34 ÷ 48

   • 1

   • 10

   • 100

3. 124.3 ÷ 20

   • 6

   • 60

   • 600

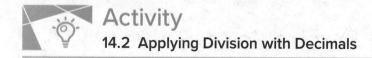

Activity

14.2 Applying Division with Decimals

Your teacher will assign to you either Problem A or Problem B. Work together as a group to answer the questions. Be prepared to create a visual display to show your reasoning with the class.

Problem A: A piece of rope is 5.75 meters in length.

1. If it is cut into 20 equal pieces, how long will each piece be?

2. If it is cut into 0.05-meter pieces, how many pieces will there be?

Problem B: A tortoise travels 0.945 miles in 3.5 hours.

1. If it moves at a constant speed, how many miles per hour is it traveling?

2. At this rate, how long will it take the tortoise to travel 4.86 miles?

NAME _____ DATE _____ PERIOD _____

Activity
14.3 Distance between Hurdles

There are 10 equally-spaced hurdles on a race track. The first hurdle is 13.72 meters from the start line. The final hurdle is 14.02 meters from the finish line. The race track is 110 meters long.

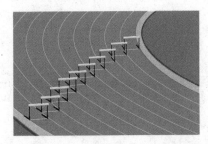

1. Draw a diagram that shows the hurdles on the race track. Label all known measurements.

2. How far are the hurdles from one another? Explain or show your reasoning.

3. A professional runner takes 3 strides between each pair of hurdles. The runner leaves the ground 2.2 meters *before* the hurdle and returns to the ground 1 meter *after* the hurdle.

   About how long are each of the runner's strides between the hurdles? Show your reasoning.

Here is a diagram of a tennis court.

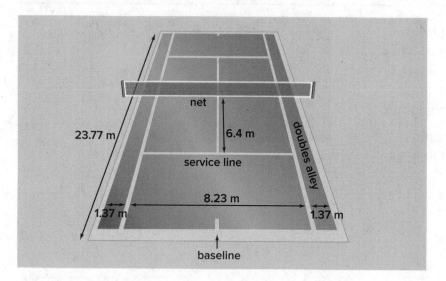

The full tennis court, used for doubles, is a rectangle. All of the angles made by the line segments in the diagram are right angles.

1. The net partitions the tennis court into two halves. Is each half a square? Explain your reasoning.

2. Is the service line halfway between the net and the baseline? Explain your reasoning.

3. Lines painted on a tennis court are 5 cm wide. A painter made markings to show the length and width of the court, then painted the lines to the outside of the markings.

   a. Did the painter's mistake increase or decrease the overall size of the tennis court? Explain how you know.

   b. By how many square meters did the court's size change? Explain your reasoning.

NAME _____ DATE _____ PERIOD _____

Summary
Using Operations on Decimals to Solve Problems

Diagrams can help us communicate and model mathematics. A clearly-labeled diagram helps us visualize what is happening in a problem and accurately communicate the information we need.

Sports offer great examples of how diagrams can help us solve problems. For example, to show the placement of the running hurdles in a diagram, we needed to know what the distances 13.72 and 14.02 meters tell us and the number of hurdles to draw. An accurate diagram not only helped us set up and solve the problem correctly, but also helped us see that there are only *nine* spaces between ten hurdles.

To communicate information clearly and solve problems correctly, it is also important to be precise in our measurements and calculations, especially when they involve decimals.

In tennis, for example, the length of the court is 23.77 meters. Because the boundary lines on a tennis court have a significant width, we would want to know whether this measurement is taken between the inside of the lines, the center of the lines, or the outside of the lines. Diagrams can help us attend to this detail, as shown here.

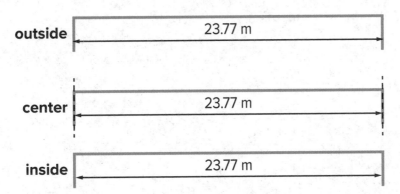

The accuracy of this measurement matters to the tennis players who use the court, so it matters to those who paint the boundaries as well. The tennis players practice their shots to be on or within certain lines. If the tennis court on which they play is not precisely measured, their shots may not land as intended in relation to the boundaries. Court painters usually need to be sure their measurements are accurate to within $\frac{1}{100}$ of a meter or one centimeter.

Using Operations on Decimals to Solve Problems

1. A roll of ribbon was 12 meters long. Diego cut 9 pieces of ribbon that were 0.4 meter each to tie some presents. He then used the remaining ribbon to make some wreaths. Each wreath required 0.6 meter. For each question, explain your reasoning.

   a. How many meters of ribbon were available for making wreaths?

   b. How many wreaths could Diego make with the available ribbon?

2. The Amazon rainforest covered 6.42 million square kilometers in 1994. In 2014, it covered only $\frac{50}{59}$ as much. Which is closest to the area of the Amazon forest in 2014? Explain how you know without calculating the exact area.

   (A.) 6.4 million km^2

   (B.) 5.4 million km^2

   (C.) 4.4 million km^2

   (D.) 3.4 million km^2

   (E.) 2.4 million km^2

NAME _____ DATE _____ PERIOD _____

3. To get an A in her math class, Jada needs to have at least 90% of the total number of points possible. The table shows Jada's results before the final test in the class.

	Jada's Points	Total Points Possible
Homework	141	150
Test 1	87	100
Test 2	81	100
Test 3	91	100

a. Does Jada have 90% of the total possible points *before* the final test? Explain how you know.

b. Jada thinks that if she gets at least 92 out of 100 on the final test, she will get an A in the class. Do you agree? Explain.

4. Find the following differences. Show your reasoning. (Lesson 5-4.)

    a. 0.151 – 0.028   b. 0.106 – 0.0315   c. 3.572 – 2.6014

5. Find these quotients. Show your reasoning. (Lesson 5-13)

    a. 24.2 ÷ 1.1   b. 13.25 ÷ 0.4   c. 170.28 ÷ 0.08

Lesson 5-15

Making and Measuring Boxes

NAME _____ DATE _____ PERIOD _____

Learning Goal Let's use what we know about decimals to make and measure boxes.

Activity
15.1 Folding Paper Boxes

Your teacher will demonstrate how to make an open-top box by folding a sheet of paper. Your group will receive 3 or more sheets of square paper. Each person in your group will make 1 box. Before you begin folding:

1. Record the side lengths of your papers, from the smallest to the largest.

   a. Paper for Box 1: _____ cm

   b. Paper for Box 2: _____ cm

   c. Paper for Box 3: _____ cm

2. Compare the side lengths of the square sheets of paper. Be prepared to explain how you know.

   a. The side length of the paper for Box 2 is _____ times the side length of the paper for Box 1.

   b. The side length of the paper for Box 3 is _____ times the side length of the paper for Box 1.

3. Make some predictions about the measurements of the three boxes your group will make:

   a. The surface area of Box 3 will be _____ as large as that of Box 1.

   b. Box 2 will be _____ times as tall as Box 1.

   c. Box 3 will be _____ times as tall as Box 1.

Now you are ready to fold your paper into a box!

Activity

15.2 Sizing Up Paper Boxes

Now that you have made your boxes, you will measure them and check your predictions about how their heights and surface areas compare.

1. a. Measure the length and height of each box to the nearest tenth of a centimeter. Record the measurements in the table.

	Side Length of Paper (cm)	Length of Box (cm)	Height of Box (cm)	Surface Area (sq cm)
Box 1				
Box 2				
Box 3				

b. Calculate the surface area of each box. Show your reasoning and decide on an appropriate level of precision for describing the surface area (Is it the nearest 10 square centimeters, nearest square centimeter, or something else?). Record your answers in the table.

2. To see how many times as large one measurement is when compared to another, we can compute their quotient. Divide each measurement of Box 2 by the corresponding measurement for Box 1 to complete the following statements.

a. The length of Box 2 is _____ times the length of Box 1.

b. The height of Box 2 is _____ times the height of Box 1.

c. The surface area of Box 2 is _____ times the surface area of Box 1.

NAME _____ DATE _____ PERIOD _____

3. Find out how the dimensions of Box 3 compare to those of Box 1 by computing quotients of their lengths, heights, and surface areas. Show your reasoning.

   a. The length of Box 3 is _____ times the length of Box 1.

   b. The height of Box 3 is _____ times the height of Box 1.

   c. The surface area of Box 3 is _____ times the surface area of Box 1.

4. Record your results in the table.

	Side Length of Paper	Length of Box	Height of Box	Surface Area
Box 2 compared to Box 1				
Box 3 compared to Box 1				

5. Earlier, in the first activity, you made predictions about how the heights and surface areas of the two larger boxes would compare to those of the smallest box. Discuss with your group:

   a. How accurate were your predictions? Were they close to the results you found by performing calculations?

   b. Let's say you had another piece of square paper to make Box 4. If the side length of this paper is 4 times the side length of the paper for Box 1, predict how the length, height, and surface area of Box 4 would compare to those of Box 1. How did you make your prediction?

Learning Targets

Lesson		Learning Target(s)
5-1	Using Decimals in a Shopping Context	• I can use decimals to make estimates and calculations about money.
5-2	Using Diagrams to Represent Addition and Subtraction	• I can use diagrams to represent and reason about addition and subtraction of decimals. • I can use place value to explain addition and subtraction of decimals. • I can use vertical calculations to represent and reason about addition and subtraction of decimals.
5-3	Adding and Subtracting Decimals with Few Non-Zero Digits	• I can tell whether writing or removing a zero in a decimal will change its value. • I know how to solve subtraction problems with decimals that require "unbundling" or "decomposing."

(continued on the next page)

(continued from the previous page)

Lesson		Learning Target(s)
5-4	Adding and Subtracting Decimals with Many Non-Zero Digits	• I can solve problems that involve addition and subtraction of decimals.
5-5	Decimal Points in Products	• I can use place value and fractions to reason about multiplication of decimals.
5-6	Methods for Multiplying Decimals	• I can use area diagrams to represent and reason about multiplication of decimals. • I know and can explain more than one way to multiply decimals using fractions and place value.
5-7	Using Diagrams to Represent Multiplication	• I can use area diagrams and partial products to represent and find products of decimals.

Lesson		Learning Target(s)
5-8	Calculating Products of Decimals	• I can describe and apply a method for multiplying decimals. • I know how to use a product of whole numbers to find a product of decimals.
5-9	Using the Partial Quotients Method	• I can use the partial quotients method to find a quotient of two whole numbers when the quotient is a whole number.
5-10	Using Long Division	• I can use long division to find a quotient of two whole numbers when the quotient is a whole number.
5-11	Dividing Numbers That Result in Decimals	• I can use long division to find the quotient of two whole numbers when the quotient is not a whole number.

(continued on the next page)

(continued from the previous page)

Lesson	Learning Target(s)
5-12 Dividing Decimals by Whole Numbers	• I can divide a decimal by a whole number. • I can explain the division of a decimal by a whole number in terms of equal-sized groups. • I know how multiplying both the dividend and the divisor by the same factor affects the quotient.
5-13 Dividing Decimals by Decimals	• I can explain how multiplying dividend and divisor by the same power of 10 can help me find a quotient of two decimals. • I can find the quotient of two decimals.
5-14 Using Operations on Decimals to Solve Problems	• I can use addition, subtraction, multiplication, and division on decimals to solve problems.
5-15 Making and Measuring Boxes	• I can use the four operations on decimals to find surface areas and reason about real-world problems.

Unit 6

Expressions and Equations

In one of the upcoming lessons, you'll apply your understanding of tables, graphs, and equations to study a boat traveling at a constant speed.

Topics

- Equations in One Variable
- Equal and Equivalent
- Expressions with Exponents
- Relationships between Quantities
- Let's Put It to Work

Unit 6

Expressions and Equations

Lesson 6-1

Tape Diagrams and Equations

NAME _____ DATE _____ PERIOD _____

Learning Goal Let's see how tape diagrams and equations can show relationships between amounts.

Warm Up
1.1 Which Diagram Is Which?

1. Here are two diagrams. One represents $2 + 5 = 7$. The other represents $5 \cdot 2 = 10$. Which is which? Label the length of each diagram.

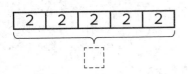

2. Draw a diagram that represents each equation.

   a. $4 + 3 = 7$

   b. $4 \cdot 3 = 12$

Activity
1.2 Match Equations and Tape Diagrams

Here are two tape diagrams. Match each equation to one of the tape diagrams.

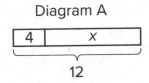

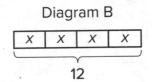

1. $4 + x = 12$

2. $12 \div 4 = x$

3. $4 \cdot x = 12$

4. $12 = 4 + x$

5. $12 - x = 4$

6. $12 = 4 \cdot x$

7. $12 - 4 = x$

8. $x = 12 - 4$

9. $x + x + x + x = 12$

For each equation, draw a diagram and find the value of the unknown that makes the equation true.

$18 = 3 + x$ $18 = 3 \cdot y$

Are you ready for more?

You are walking down a road, seeking treasure. The road branches off into three paths. A guard stands in each path. You know that only one of the guards is telling the truth, and the other two are lying. Here is what they say:

- Guard 1: The treasure lies down this path.

- Guard 2: No treasure lies down this path; seek elsewhere.

- Guard 3: The first guard is lying.

Which path leads to the treasure?

NAME _____ DATE _____ PERIOD _____

Summary
Tape Diagrams and Equations

Tape diagrams can help us understand relationships between quantities and how operations describe those relationships.

Diagram A

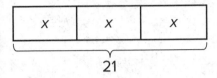

Diagram A has 3 parts that add to 21. Each part is labeled with the same letter, so we know the three parts are equal.

Here are some equations that all represent Diagram A.

$x + x + x = 21$

$3 \cdot x = 21$

$x = 21 \div 3$

$x = \frac{1}{3} \cdot 21$

Notice that the number 3 is not seen in the diagram; the 3 comes from counting 3 boxes representing 3 equal parts in 21.

We can use the diagram or any of the equations to reason that the value of x is 7.

Diagram B

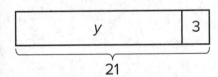

Diagram B has 2 parts that add to 21.

Here are some equations that all represent Diagram B.

$y + 3 = 21$

$y = 21 - 3$

$3 = 21 - y$

We can use the diagram or any of the equations to reason that the value of y is 18.

1. Here is an equation: $x + 4 = 17$

    a. Draw a tape diagram to represent the equation.

    b. Which part of the diagram shows the quantity x? What about 4? What about 17?

    c. How does the diagram show that $x + 4$ has the same value as 17?

2. Diego is trying to find the value of x in $5 \cdot x = 35$. He draws this diagram but is not certain how to proceed.

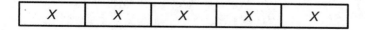

    a. Complete the tape diagram so it represents the equation $5 \cdot x = 35$.
    b. Find the value of x.

NAME _____ DATE _____ PERIOD _____

3. Match each equation to one of the two tape diagrams.

Equations

   a. $x + 3 = 9$

   b. $3 \cdot x = 9$

   c. $9 = 3 \cdot x$

   d. $3 + x = 9$

   e. $x = 9 - 3$

   f. $x = 9 \div 3$

   g. $x + x + x = 9$

Diagrams

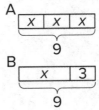

4. For each equation, draw a tape diagram and find the unknown value.

   a. $x + 9 = 16$

   b. $4 \cdot x = 28$

5. A shopper paid $2.52 for 4.5 pounds of potatoes, $7.75 for 2.5 pounds of broccoli, and $2.45 for 2.5 pounds of pears. What is the unit price of each item she bought? Show your reasoning. **(Lesson 5-13)**

6. A sports drink bottle contains 16.9 fluid ounces. Andre drank
   80% of the bottle. How many fluid ounces did Andre drink?
   Show your reasoning. (Lesson 3-14)

7. The daily recommended allowance of calcium for a sixth grader is
   1,200 mg. One cup of milk has 25% of the recommended daily allowance of
   calcium. How many milligrams of calcium are in a cup of milk? If you get
   stuck, consider using the double number line. (Lesson 3-11)

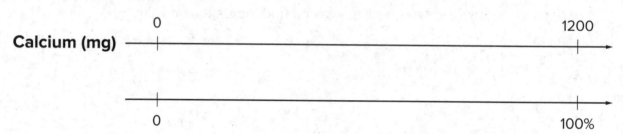

Lesson 6-2

Truth and Equations

NAME _____ DATE _____ PERIOD _____

Learning Goal Let's use equations to represent stories and see what it means to solve equations.

 ## Warm Up
2.1 Three Letters

1. The equation $a + b = c$ could be true or false.

   a. If a is 3, b is 4, and c is 5, is the equation true or false?

   b. Find new values of a, b, and c that make the equation true.

   c. Find new values of a, b, and c that make the equation false.

2. The equation $x \cdot y = z$ could be true or false.

   a. If x is 3, y is 4, and z is 12, is the equation true or false?

   b. Find new values of x, y, and z that make the equation true.

   c. Find new values of x, y, and z that make the equation false.

Activity

2.2 Storytime

Here are three situations and six equations. Which equation best represents each situation? If you get stuck, consider drawing a diagram.

$$x + 5 = 20 \qquad x + 20 = 5 \qquad x = 20 + 5$$

$$5 \cdot 20 = x \qquad 5x = 20 \qquad 20x = 5$$

1. After Elena ran 5 miles on Friday, she had run a total of 20 miles for the week. She ran x miles before Friday.

2. Andre's school has 20 clubs, which is five times as many as his cousin's school. His cousin's school has x clubs.

3. Jada volunteers at the animal shelter. She divided 5 cups of cat food equally to feed 20 cats. Each cat received x cups of food.

NAME _____ DATE _____ PERIOD _____

Activity

2.3 Using Structure to Find Solutions

Here are some equations that contain a **variable** and a list of values. Think about what each equation means and find a **solution** in the list of values. If you get stuck, consider drawing a diagram. Be prepared to explain why your solution is correct.

List: $\dfrac{1}{8}$ $\dfrac{3}{7}$ $\dfrac{4}{7}$ $\dfrac{3}{5}$ $\dfrac{5}{3}$ $\dfrac{7}{3}$ 0.01 0.1 0.5

.1 2 8.5 9.5 16.7 20 400 600 1400

1. $1000 - a = 400$

2. $12.6 = b + 4.1$

3. $8c = 8$

4. $\dfrac{2}{3} \cdot d = \dfrac{10}{9}$

5. $10e = 1$

6. $10 = 0.5f$

7. $0.99 = 1 - g$

8. $h + \dfrac{3}{7} = 1$

Are you ready for more?

One solution to the equation $a + b + c = 10$ is $a = 2$, $b = 5$, $c = 3$.

How many different whole-number solutions are there to the equation $a + b + c = 10$? Explain or show your reasoning.

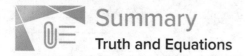

Summary
Truth and Equations

An equation can be true or false.

- An example of a true equation is $7 + 1 = 4 \cdot 2$.

- An example of a false equation is $7 + 1 = 9$.

An equation can have a letter in it, for example, $u + 1 = 8$.

- This equation is false if u is 3, because $3 + 1$ does not equal 8.

- This equation is true if u is 7, because $7 + 1 = 8$.

A letter in an equation is called a **variable**. In $u + 1 = 8$, the variable is u. A number that can be used in place of the variable that makes the equation true is called a **solution** to the equation. In $u + 1 = 8$, the solution is 7.

When a number is written next to a variable, the number and the variable are being multiplied.

- For example, $7x = 21$ means the same thing as $7 \cdot x = 21$.

A number written next to a variable is called a **coefficient**. If no coefficient is written, the coefficient is 1.

- For example, in the equation $p + 3 = 5$, the coefficient of p is 1.

Glossary

coefficient

solution to an equation

variable

NAME _____ DATE _____ PERIOD _____

Practice
Truth and Equations

1. Select **all** the true equations.

(A.) $5 + 0 = 0$

(B.) $15 \cdot 0 = 0$

(C.) $1.4 + 2.7 = 4.1$

(D.) $\frac{2}{3} \cdot \frac{5}{9} = \frac{7}{12}$

(E.) $4\frac{2}{3} = 5 - \frac{1}{3}$

2. Mai's water bottle had 24 ounces in it. After she drank x ounces of water, there were 10 ounces left. Select **all** the equations that represent this situation.

(A.) $24 \div 10 = x$

(B.) $24 + 10 = x$

(C.) $24 - 10 = x$

(D.) $x + 10 = 24$

(E.) $10x = 24$

3. Priya has 5 pencils, each x inches in length. When she lines up the pencils end to end, they measure 34.5 inches. Select **all** the equations that represent this situation.

(A.) $5 + x = 34.5$

(B.) $5x = 34.5$

(C.) $34.5 \div 5 = x$

(D.) $34.5 - 5 = x$

(E.) $x = (34.5) \cdot 5$

4. Match each equation with a solution from the list of values.

a. $2a = 4.6$

b. $b + 2 = 4.6$

c. $c \div 2 = 4.6$

d. $d - 2 = 4.6$

e. $e + \frac{3}{8} = 2$

f. $\frac{1}{8}f = 3$

g. $g \div \frac{8}{5} = 1$

$\frac{8}{5}$

$1\frac{5}{8}$

2.3

2.6

6.6

9.2

24

5. The daily recommended allowance of vitamin C for a sixth grader is 45 mg. 1 orange has about 75% of the recommended daily allowance of vitamin C. How many milligrams are in 1 orange? If you get stuck, consider using the double number line. (Lesson 3-11)

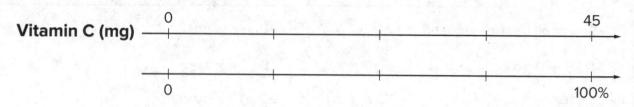

6. There are 90 kids in the band. 20% of the kids own their own instruments, and the rest rent them. (Lesson 3-12)

    a. How many kids own their own instruments?

    b. How many kids rent instruments?

    c. What percentage of kids rent their instruments?

7. Find each product. (Lesson 5-8)

    a. $(0.25) \cdot (1.4)$

    b. $(0.061) \cdot (0.43)$

    c. $(1.017) \cdot (0.072)$

    d. $(5.226) \cdot (0.037)$

Lesson 6-3

Staying in Balance

NAME _____ DATE _____ PERIOD _____

Learning Goal Let's use balanced hangers to help us solve equations.

Warm Up
3.1 Hanging Around

1. For diagram A, find:

   a. one thing that *must* be true.

   b. one thing that *could* be true or false.

   c. one thing that *cannot possibly* be true.

Diagram A

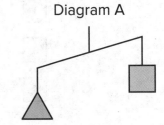

2. For diagram B, find:

   a. one thing that *must* be true.

   b. one thing that *could* be true or false.

   c. one thing that *cannot possibly* be true.

Diagram B

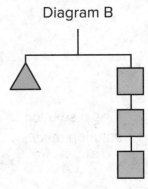

3.2 Match Equations and Hangers

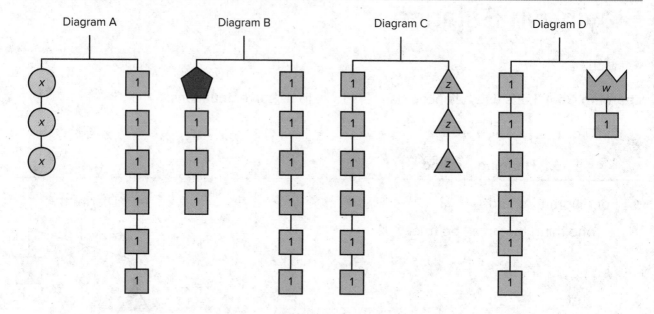

1. Match each hanger to an equation. Complete the equation by writing x, y, z, or w in the empty box.

   $\boxed{} + 3 = 6$ $3 \cdot \boxed{} = 6$ $6 = \boxed{} + 1$ $6 = 3 \cdot \boxed{}$

2. Find a solution to each equation. Use the hangers to explain what each solution means.

NAME _____ DATE _____ PERIOD _____

 ## Activity
3.3 Connecting Diagrams to Equations and Solutions

Here are some balanced hangers. Each piece is labeled with its weight.

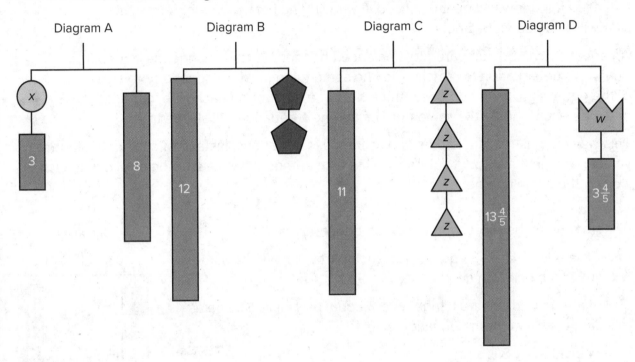

For each diagram:

1. Write an equation.

2. Explain how to reason with the *diagram* to find the weight of a piece with a letter.

3. Explain how to reason with the *equation* to find the weight of a piece with a letter.

When you have the time, go online to solve some trickier puzzles that use hanger diagrams like the ones in this lesson. You can even build new ones. (If you want to do this during class, check with your teacher first!)

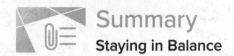

Summary
Staying in Balance

A hanger stays balanced when the weights on both sides are equal. We can change the weights and the hanger will stay balanced as long as both sides are changed in the same way. For example, adding 2 pounds to each side of a balanced hanger will keep it balanced. Removing half of the weight from each side will also keep it balanced.

An equation can be compared to a balanced hanger. We can change the equation, but for a true equation to remain true, the same thing must be done to both sides of the equal sign. If we add or subtract the same number on each side, or multiply or divide each side by the same number, the new equation will still be true.

This way of thinking can help us find solutions to equations. Instead of checking different values, we can think about subtracting the same amount from each side or dividing each side by the same number.

Diagram A can be represented by the equation $3x = 11$.

If we break the 11 into 3 equal parts, each part will have the same weight as a block with an x.

Splitting each side of the hanger into 3 equal parts is the same as dividing each side of the equation by 3.

- $3x$ divided by 3 is x.

- 11 divided by 3 is $\frac{11}{3}$.

- If $3x = 11$ is true, then $x = \frac{11}{3}$ is true.

- The solution to $3x = 11$ is $\frac{11}{3}$.

Diagram A

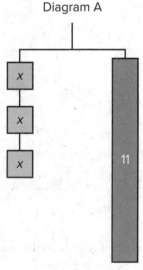

Diagram B can be represented with the equation $11 = y + 5$.

If we remove a weight of 5 from each side of the hanger, it will stay in balance.

Removing 5 from each side of the hanger is the same as subtracting 5 from each side of the equation.

- $11 - 5$ is 6.

- $y + 5 - 5$ is y.

- If $11 = y + 5$ is true, then $6 = y$ is true.

- The solution to $11 = y + 5$ is 6.

Diagram B

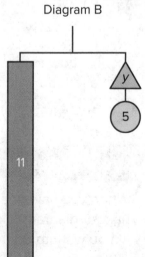

NAME _____ DATE _____ PERIOD _____

Practice
Staying in Balance

1. Select **all** the equations that represent the hanger.

(A.) $x + x + x = 1 + 1 + 1 + 1 + 1 + 1$

(B.) $x \cdot x \cdot x = 6$

(C.) $3x = 6$

(D.) $x + 3 = 6$

(E.) $x \cdot x \cdot x = 1 \cdot 1 \cdot 1 \cdot 1 \cdot 1 \cdot 1$

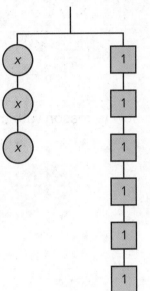

2. Write an equation to represent each hanger.

a.   b.   c.   d.

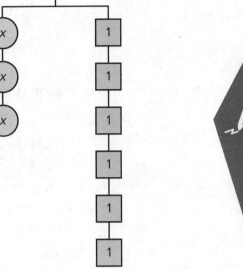

3. Respond to each of the following.

   a. Write an equation to represent the hanger.

   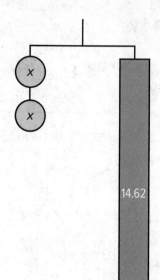

   b. Explain how to reason with the hanger to find the value of x.

   c. Explain how to reason with the equation to find the value of x.

NAME _____ DATE _____ PERIOD _____

4. Andre says that *x* is 7 because he can move the two 1s with the *x* to the other side. Do you agree with Andre? Explain your reasoning.

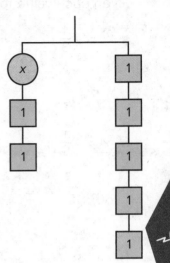

5. Match each equation to one of the diagrams. **(Lesson 6-1)**

A
m	

12	4

B
12	

4	*m*

C
m			

12	12	12	12

D
12			

m	*m*	*m*	*m*

a. $12 - m = 4$

b. $12 = 4 \cdot m$

c. $m - 4 = 12$

d. $\frac{m}{4} = 12$

6. The area of a rectangle is 14 square units. It has side lengths x and y. Given each value for x, find y. (Lesson 4-13)

   a. $x = 2\frac{1}{3}$

   b. $x = 4\frac{1}{5}$

   c. $x = \frac{7}{6}$

7. Lin needs to save up $20 for a new game. How much money does she have if she has saved each percentage of her goal? Explain your reasoning. (Lesson 3-11)

   a. 25%

   b. 75%

   c. 125%

Lesson 6-4

Practice Solving Equations and Representing Situations with Equations

NAME _____ DATE _____ PERIOD _____

Learning Goal Let's solve equations by doing the same to each side.

Warm Up

4.1 Number Talk: Subtracting From Five

Find the value of each expression mentally.

1. $5 - 2$ **2.** $5 - 2.1$ **3.** $5 - 2.17$ **4.** $5 - 2\frac{7}{8}$

Activity

4.2 Row Game: Solving Equations Practice

Solve the equations in one column. Your partner will work on the other column.

Check in with your partner after you finish each row. Your answers in each row should be the same. If your answers aren't the same, work together to find the error and correct it.

Row	Column A	Column B
1	$18 = 2x$	$36 = 4x$
2	$17 = x + 9$	$13 = x + 5$
3	$8x = 56$	$3x = 21$
4	$21 = \frac{1}{4}x$	$28 = \frac{1}{3}x$
5	$6x = 45$	$8x = 60$
6	$x + 4\frac{5}{6} = 9$	$x + 3\frac{5}{6} = 8$
7	$\frac{5}{7}x = 55$	$\frac{3}{7}x = 33$
8	$\frac{1}{5} = 6x$	$\frac{1}{3} = 10x$
9	$2.17 + x = 5$	$6.17 + x = 9$
10	$\frac{20}{3} = \frac{10}{9}x$	$\frac{14}{5} = \frac{7}{15}x$
11	$14.88 + x = 17.05$	$3.91 + x = 6.08$
12	$3\frac{3}{4}x = 1\frac{1}{4}$	$\frac{7}{5}x = \frac{7}{15}$

Activity

4.3 Choosing Equations to Match Situations

Circle **all** of the equations that describe each situation. If you get stuck, consider drawing a diagram. Then find the solution for each situation.

1. Clare has 8 fewer books than Mai. If Mai has 26 books, how many books does Clare have?

   $26 - x = 8$

   $x = 26 + 8$

   $x + 8 = 26$

   $26 - 8 = x$

   $x =$ _____

2. A coach formed teams of 8 from all the players in a soccer league. There are 14 teams. How many players are in the league?

   $y = 14 \div 8$

   $\dfrac{y}{8} = 14$

   $\dfrac{1}{8}y = 14$

   $y = 14 \cdot 8$

   $y =$ _____

3. Kiran scored 223 more points in a computer game than Tyler. If Kiran scored 409 points, how many points did Tyler score?

   $223 = 409 - z$

   $409 - 223 = z$

   $409 + 223 = z$

   $409 = 223 + z$

   $z =$ _____

4. Mai ran 27 miles last week, which was three times as far as Jada ran. How far did Jada run?

   $3w = 27$

   $w = \dfrac{1}{3} \cdot 27$

   $w = 27 \div 3$

   $w = 3 \cdot 27$

   $w =$ _____

NAME _____ DATE _____ PERIOD _____

Mai's mother was 28 when Mai was born. Mai is now 12 years old. In how many years will Mai's mother be twice Mai's age? How old will they be then?

Summary
Practice Solving Equations and Representing Situations with Equations

Writing and solving equations can help us answer questions about situations.

Suppose a scientist has 13.68 liters of acid and needs 16.05 liters for an experiment. How many more liters of acid does she need for the experiment?

- We can represent this situation with the equation:

  $13.68 + x = 16.05$

- When working with hangers, we saw that the solution can be found by subtracting 13.68 from each side. This gives us some new equations that also represent the situation:

  $x = 16.05 - 13.68$

  $x = 2.37$

- Finding a solution in this way leads to a variable on one side of the equal sign and a number on the other. We can easily read the solution—in this case, 2.37—from an equation with a letter on one side and a number on the other. We often write solutions in this way.

Let's say a food pantry takes a 54-pound bag of rice and splits it into portions that each weigh $\frac{3}{4}$ of a pound. How many portions can they make from this bag?

- We can represent this situation with the equation:

  $\frac{3}{4}x = 54$

- We can find the value of x by dividing each side by $\frac{3}{4}$. This gives us some new equations that represent the same situation:

  $x = 54 \div \frac{3}{4}$

  $x = 72$

- The solution is 72 portions.

Practice

Practice Solving Equations and Representing Situations with Equations

1. Select **all** the equations that describe each situation and then find the solution.

   a. Kiran's backpack weighs 3 pounds less than Clare's backpack. Clare's backpack weighs 14 pounds. How much does Kiran's backpack weigh?

   $x + 3 = 14$ $\qquad$ $3x = 14$ $\qquad$ $x = 14 - 3$ $\qquad$ $x = 14 \div 3$

   b. Each notebook contains 60 sheets of paper. Andre has 5 notebooks. How many sheets of paper do Andre's notebooks contain?

   $y = 60 \div 5$ $\qquad$ $y = 5 \cdot 60$ $\qquad$ $\frac{y}{5} = 60$ $\qquad$ $5y = 60$

2. Solve each equation.

   a. $2x = 5$

   b. $y + 1.8 = 14.7$

   c. $6 = \frac{1}{2}z$

   d. $3\frac{1}{4} = \frac{1}{2} + w$

   e. $2.5t = 10$

NAME _____ DATE _____ PERIOD _____

3. For each equation, draw a tape diagram that represents the equation. (Lesson 6-1)

   a. $3 \cdot x = 18$

   b. $3 + x = 18$

   c. $17 - 6 = x$

4. Find each product. (Lesson 5-8)

   a. $(21.2) \cdot (0.02)$

   b. $(2.05) \cdot (0.004)$

5. For a science experiment, students need to find 25% of 60 grams.

   - Jada says, "I can find this by calculating $\frac{1}{4}$ of 60."
   - Andre says, "25% of 60 means $\frac{25}{100} \cdot 60$."

   Do you agree with either of them? Explain your reasoning. (Lesson 3-13)

Lesson 6-5

A New Way to Interpret *a* over *b*

NAME _____ DATE _____ PERIOD _____

Learning Goal Let's investigate what a fraction means when the numerator and denominator are not whole numbers.

Warm Up
5.1 Recalling Ways of Solving

Solve each equation. Be prepared to explain your reasoning.

1. $0.07 = 10m$ **2.** $10.1 = t + 7.2$

Activity
5.2 Interpreting $\frac{a}{b}$

Solve each equation.

1. $35 = 7x$ **2.** $35 = 11x$

3. $7x = 7.7$ **4.** $0.3x = 2.1$

5. $\frac{2}{5} = \frac{1}{2}x$

Solve the equation. Try to find some shortcuts.

$$\frac{1}{6} \cdot \frac{3}{20} \cdot \frac{5}{42} \cdot \frac{7}{72} \cdot x = \frac{1}{384}$$

Activity

5.3 Storytime Again

Take turns with your partner telling a story that might be represented by each equation. Then, for each equation, choose one story, state what quantity x describes, and solve the equation. If you get stuck, consider drawing a diagram.

1. $0.7 + x = 12$

2. $\frac{1}{4}x = \frac{3}{2}$

NAME _____ DATE _____ PERIOD _____

Summary
A New Way to Interpret *a* over *b*

In the past, you learned that a fraction such as $\frac{4}{5}$ can be thought of in a few ways.

- $\frac{4}{5}$ is a number you can locate on the number line by dividing the section between 0 and 1 into 5 equal parts and then counting 4 of those parts to the right of 0.

- $\frac{4}{5}$ is the share that each person would have if 4 wholes were shared equally among 5 people. This means that $\frac{4}{5}$ is the result of *dividing* 4 by 5.

We can extend this meaning of *a fraction as a quotient* to fractions whose numerators and denominators are not whole numbers. For example, we can represent 4.5 pounds of rice divided into portions that each weigh 1.5 pounds as: $\frac{4.5}{1.5} = 4.5 \div 1.5 = 3$. In other words, $\frac{4.5}{1.5} = 3$ because the quotient of 4.5 and 1.5 is 3.

Fractions that involve non-whole numbers can also be used when we solve equations.

Suppose a road under construction is $\frac{3}{8}$ finished and the length of the completed part is $\frac{4}{3}$ miles. How long will the road be when completed?

We can write the equation $\frac{3}{8}x = \frac{4}{3}$ to represent the situation and solve the equation.

$$\frac{3}{8}x = \frac{4}{3}$$

$$x = \frac{\frac{4}{3}}{\frac{3}{8}}$$

$$x = \frac{4}{3} \cdot \frac{8}{3}$$

$$x = \frac{32}{9} \text{ or } 3\frac{5}{9}$$

The completed road will be $3\frac{5}{9}$ or about 3.6 miles long.

1. Select **all** the expressions that equal $\frac{3.15}{0.45}$.

   (A.) $(3.15) \cdot (0.45)$

   (B.) $(3.15) \div (0.45)$

   (C.) $(3.15) \cdot \frac{1}{0.45}$

   (D.) $(3.15) \div \frac{45}{100}$

   (E.) $(3.15) \cdot \frac{100}{45}$

   (F.) $\frac{0.45}{3.15}$

2. Which expressions are solutions to the equation $\frac{3}{4}x = 15$?
   Select **all** that apply.

   (A.) $\dfrac{15}{\frac{3}{4}}$

   (B.) $\dfrac{15}{\frac{4}{3}}$

   (C.) $\frac{4}{3} \cdot 15$

   (D.) $\frac{3}{4} \cdot 15$

   (E.) $15 \div \frac{3}{4}$

3. Solve each equation.

   a. $4a = 32$

   b. $4 = 32b$

   c. $10c = 26$

   d. $26 = 100d$

NAME _____ DATE _____ PERIOD _____

4. For each equation, write a story problem represented by the equation. For each equation, state what quantity x represents. If you get stuck, consider drawing a diagram.

a. $\frac{3}{4} + x = 2$

b. $1.5x = 6$

5. Write as many mathematical expressions or equations as you can about the image. Include a fraction, a decimal number, or a percentage in each. **(Lesson 3-13)**

FUNDRAISER

Our Goal
$250,000

$250,000

$200,000

$150,000

$100,000

$50,000

6. In a lilac paint mixture, 40% of the mixture is white paint, 20% is blue, and the rest is red. There are 4 cups of blue paint used in a batch of lilac paint. (Lesson 3-12)

   a. How many cups of white paint are used?

   b. How many cups of red paint are used?

   c. How many cups of lilac paint will this batch yield?

   If you get stuck, consider using a tape diagram.

7. Triangle P has a base of 12 inches and a corresponding height of 8 inches. Triangle Q has a base of 15 inches and a corresponding height of 6.5 inches. Which triangle has a greater area? Show your reasoning.
   (Lesson 1-9)

Lesson 6-6

Write Expressions Where Letters Stand for Numbers

NAME _____ DATE _____ PERIOD _____

Learning Goal Let's use expressions with variables to describe situations.

 ## Warm Up
6.1 Algebra Talk: When x Is 6

If x is 6, what is:

a. $x + 4$

b. $7 - x$

c. x^2

d. $\frac{1}{3}x$

 ## Activity
6.2 Lemonade Sales and Heights

1. Lin set up a lemonade stand. She sells the lemonade for $0.50 per cup.

   a. Complete the table to show how much money she would collect if she sold each number of cups.

Lemonade Sold (number of cups)	12	183	c
Money Collected (dollars)			

   b. How many cups did she sell if she collected $127.50? Be prepared to explain your reasoning.

2. Elena is 59 inches tall. Some other people are taller than Elena.

   a. Complete the table to show the height of each person.

Person	Andre	Lin	Noah
How Much Taller Than Elena (inches)	4	$6\frac{1}{2}$	d
Person's Height (inches)			

   b. If Noah is $64\frac{3}{4}$ inches tall, how much taller is he than Elena?

Activity

6.3 Building Expressions

1. Clare is 5 years older than her cousin.

   a. How old would Clare be if her cousin is:

   • 10 years old?

   • 2 years old?

   • x years old?

   b. Clare is 12 years old. How old is Clare's cousin?

2. Diego has 3 times as many comic books as Han.

   a. How many comic books does Diego have if Han has:

   • 6 comic books?

   • n books?

   b. Diego has 27 comic books. How many comic books does Han have?

NAME _____ DATE _____ PERIOD _____

3. Two-fifths of the vegetables in Priya's garden are tomatoes.

   a. How many tomatoes are there if Priya's garden has:

   • 20 vegetables?

   • *x* vegetables?

   b. Priya's garden has 6 tomatoes. How many total vegetables are there?

4. A school paid $31.25 for each calculator.

   a. If the school bought *x* calculators, how much did they pay?

   b. The school spent $500 on calculators. How many did the school buy?

Are you ready for more?

Kiran, Mai, Jada, and Tyler went to their school carnival. They all won chips that they could exchange for prizes. Kiran won $\frac{2}{3}$ as many chips as Jada. Mai won 4 times as many chips as Kiran. Tyler won half as many chips as Mai.

1. Write an expression for the number of chips Tyler won. You should only use one variable, *J*, which stands for the number of chips Jada won.

2. If Jada won 42 chips, how many chips did Tyler, Kiran, and Mai each win?

Suppose you share a birthday with a neighbor, but she is 3 years older than you.

- When you were 1, she was 4.

- When you were 9, she was 12.

- When you are 42, she will be 45.

If we let a represent your age at any time, your neighbor's age can be expressed $a + 3$.

Your Age	1	9	42	a
Neighbor's Age	4	12	45	$a + 3$

We often use a letter such as x or a as a placeholder for a number in expressions. These are called *variables* (just like the letters we used in equations, previously). Variables make it possible to write expressions that represent a calculation even when we don't know all the numbers in the calculation.

How old will you be when your neighbor is 32?

Since your neighbor's age is calculated with the expression $a + 3$, we can write the equation $a + 3 = 32$.

When your neighbor is 32 you will be 29, because $a + 3 = 32$ is true when a is 29.

NAME _____ DATE _____ PERIOD _____

Practice
Write Expressions Where Letters Stand for Numbers

1. Instructions for a craft project say that the length of a piece of red ribbon should be 7 inches less than the length of a piece of blue ribbon.

   a. How long is the red ribbon if the length of the blue ribbon is:

   • 10 inches?

   • 27 inches?

   • x inches?

   b. How long is the blue ribbon if the red ribbon is 12 inches?

2. Tyler has 3 times as many books as Mai.

   a. How many books does Mai have if Tyler has:

   • 15 books?

   • 21 books?

   • x books?

   b. Tyler has 18 books. How many books does Mai have?

3. A bottle holds 24 ounces of water. It has x ounces of water in it.

   a. What does $24 - x$ represent in this situation?

   b. Write a question about this situation that has $24 - x$ for the answer.

4. Write an equation represented by this tape diagram using these operations. **(Lesson 6-1)**

   a. addition

   b. subtraction

   c. multiplication

   d. division

5. Select **all** the equations that describe each situation. Then find the solution. (Lesson 6-4)

   a. Han's house is 450 meters from school. Lin's house is 135 meters closer to school. How far is Lin's house from school?

   $z = 450 + 135$ $z = 450 - 135$ $z - 135 = 450$ $z + 135 = 450$

   b. Tyler's playlist has 36 songs. Noah's playlist has one quarter as many songs as Tyler's playlist. How many songs are on Noah's playlist?

   $w = 4 \cdot 36$ $w = 36 \div 4$ $4w = 36$ $\dfrac{w}{4} = 36$

6. You had $50. You spent 10% of the money on clothes, 20% on games, and the rest on books. How much money was spent on books? (Lesson 3-12)

7. A trash bin has a capacity of 50 gallons. What percentage of its capacity is each amount? Show your reasoning. (Lesson 3-14)

   a. 5 gallons

   b. 30 gallons

   c. 45 gallons

   d. 100 gallons

Lesson 6-7

Revisit Percentages

NAME _____ DATE _____ PERIOD _____

Learning Goal Let's use equations to find percentages.

Warm Up
7.1 Number Talk: Percentages

Solve each problem mentally.

1. Bottle A contains 4 ounces of water, which is 25% of the amount of water in Bottle B. How much water is there in Bottle B?

2. Bottle C contains 150% of the water in Bottle B. How much water is there in Bottle C?

3. Bottle D contains 12 ounces of water. What percentage of the amount of water in Bottle B is this?

Activity
7.2 Representing a Percentage Problem with an Equation

1. Answer each question and show your reasoning.

   a. Is 60% of 400 equal to 87?

   b. Is 60% of 200 equal to 87?

   c. Is 60% of 120 equal to 87?

2. 60% of x is equal to 87. Write an equation that expresses the relationship between 60%, x, and 87. Solve your equation.

3. Write an equation to help you find the value of each variable. Solve the equation.

   a. 60% of c is 43.2.

   b. 38% of e is 190.

Activity

7.3 Puppies Grow Up, Revisited

1. Puppy A weighs 8 pounds, which is about 25% of its adult weight. What will be the adult weight of Puppy A?

2. Puppy B weighs 8 pounds, which is about 75% of its adult weight. What will be the adult weight of Puppy B?

3. If you haven't already, write an equation for each situation. Then, show how you could find the adult weight of each puppy by solving the equation.

NAME _____ DATE _____ PERIOD _____

Are you ready for more?

Diego wants to paint his room purple. He bought one gallon of purple paint that is 30% red paint and 70% blue paint. Diego wants to add more blue to the mix so that the paint mixture is 20% red, 80% blue.

1. How much blue paint should Diego add? Test the following possibilities: 0.2 gallons, 0.3 gallons, 0.4 gallons, 0.5 gallons.

2. Write an equation in which x represents the amount of paint Diego should add.

3. Check that the amount of paint Diego should add is a solution to your equation.

Summary
Revisit Percentages

If we know that 455 students are in school today and that number represents 70% attendance, we can write an equation to figure out how many students go to the school.

The number of students in school today is known in two different ways: as 70% of the students in the school, and also as 455. If s represents the total number of students who go to the school, then 70% of s, or $\frac{70}{100}s$, represents the number of students that are in school today, which is 455.

We can write and solve the equation:

$$\frac{70}{100}s = 455$$

$$s = 455 \div \frac{70}{100}$$

$$s = 455 \cdot \frac{100}{70}$$

$$s = 650$$

There are 650 students in the school.

In general, equations can help us solve problems in which one amount is a percentage of another amount.

Practice
Revisit Percentages

1. A crew has paved $\frac{3}{4}$ of a mile of road. If they have completed 50% of the work, how long is the road they are paving?

2. 40% of x is 35.

   a. Write an equation that shows the relationship of 40%, x, and 35.

   b. Use your equation to find x. Show your reasoning.

3. Priya has completed 9 exam questions. This is 60% of the questions on the exam.

   a. Write an equation representing this situation. Explain the meaning of any variables you use.

   b. How many questions are on the exam? Show your reasoning.

NAME _____ DATE _____ PERIOD _____

4. Answer each question. Show your reasoning.

   a. 20% of a is 11. What is a?

   b. 75% of b is 12. What is b?

   c. 80% of c is 20. What is c?

   d. 200% of d is 18. What is d?

5. For the equation $2n - 3 = 7$: (Lesson 6-2)

    a. What is the variable?

    b. What is the coefficient of the variable?

    c. Which of these is the solution to the equation?

       2, 3, 5, 7, n

6. Which of these is a solution to the equation $\frac{1}{8} = \frac{2}{5} \cdot x$? (Lesson 6-2)

    Ⓐ $\frac{2}{40}$

    Ⓑ $\frac{5}{16}$

    Ⓒ $\frac{11}{40}$

    Ⓓ $\frac{17}{40}$

7. Find the quotients. (Lesson 5-13)

    a. $0.009 \div 0.001$

    b. $0.009 \div 0.002$

    c. $0.0045 \div 0.001$

    d. $0.0045 \div 0.002$

Lesson 6-8

Equal and Equivalent

NAME _____ DATE _____ PERIOD _____

Learning Goal Let's use diagrams to figure out which expressions are equivalent and which are just sometimes equal.

Warm Up
8.1 Algebra Talk: Solving Equations by Seeing Structure

Find a solution to each equation mentally.

1. $3 + x = 8$

2. $10 = 12 - x$

3. $x^2 = 49$

4. $\frac{1}{3}x = 6$

Here is a diagram of $x + 2$ and $3x$ when x is 4. Notice that the two diagrams are lined up on their left sides.

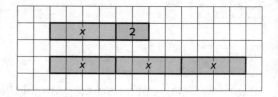

In each of your drawings below, line up the diagrams on one side.

1. Draw a diagram of $x + 2$, and a separate diagram of $3x$, when x is 3.

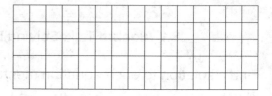

2. Draw a diagram of $x + 2$, and a separate diagram of $3x$, when x is 2.

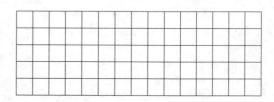

3. Draw a diagram of $x + 2$, and a separate diagram of $3x$, when x is 1.

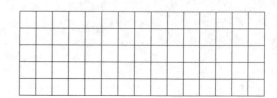

4. Draw a diagram of $x + 2$, and a separate diagram of $3x$, when x is 0.

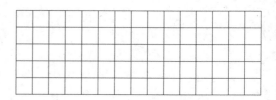

NAME _____ DATE _____ PERIOD _____

5. When are $x + 2$ and $3x$ equal? When are they not equal? Use your diagrams to explain.

6. Draw a diagram of $x + 3$, and a separate diagram of $3 + x$.

7. When are $x + 3$ and $3 + x$ equal? When are they not equal? Use your diagrams to explain.

Here is a list of expressions. Find any pairs of expressions that are equivalent.
If you get stuck, try reasoning with diagrams.

$a + 3$ $a + a + a$

$a \div \dfrac{1}{3}$ $a \cdot 3$

$\dfrac{1}{3}a$ $3a$

$\dfrac{a}{3}$ $1a$

a $3 + a$

Are you ready for more?

Below are four questions about equivalent expressions. For each one:

- Decide whether you think the expressions are equivalent.

- Test your guess by choosing numbers for x (and y, if needed).

1. Are $\dfrac{x \cdot x \cdot x \cdot x}{x}$ and $x \cdot x \cdot x$ equivalent expressions?

2. Are $\dfrac{x + x + x + x}{x}$ and $x + x + x$ equivalent expressions?

3. Are $2(x + y)$ and $2x + 2y$ equivalent expressions?

4. Are $2xy$ and $2x \cdot 2y$ equivalent expressions?

NAME _____ DATE _____ PERIOD _____

Summary
Equal and Equivalent

We can use diagrams showing lengths of rectangles to see when expressions are equal.

For example, the expressions $x + 9$ and $4x$ are equal when x is 3, but are not equal for other values of x.

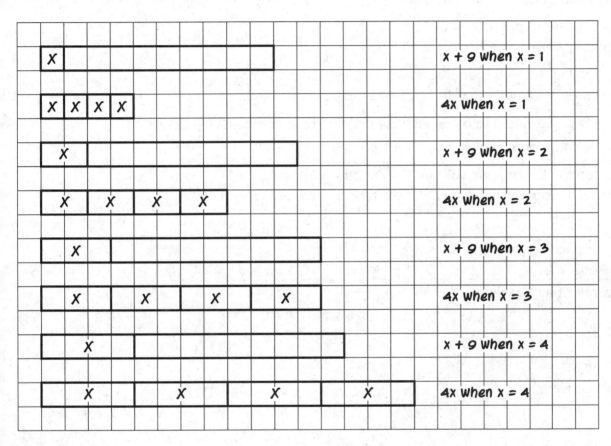

Sometimes two expressions are equal for only one particular value of their variable.

Other times, they seem to be equal no matter what the value of the variable.

Expressions that are always equal for the same value of their variable are called **equivalent expressions**. However, it would be impossible to test every possible value of the variable. How can we know for sure that expressions are equivalent?

We use the meaning of operations and properties of operations to know that expressions are equivalent.

Here are some examples:

- $x + 3$ is equivalent to $3 + x$ because of the commutative property of addition.

- $4 \cdot y$ is equivalent to $y \cdot 4$ because of the commutative property of multiplication.

- $a + a + a + a + a$ is equivalent to $5 \cdot a$ because adding 5 copies of something is the same as multiplying it by 5.

- $b \div 3$ is equivalent to $b \cdot \frac{1}{3}$ because dividing by a number is the same as multiplying by its reciprocal.

In the coming lessons, we will see how another property, the distributive property, can show that expressions are equivalent.

Glossary
_____

equivalent expressions

NAME _____ DATE _____ PERIOD _____

Practice
Equal and Equivalent

1. **a.** Draw a diagram of $x + 3$ and a diagram of $2x$ when x is 1.

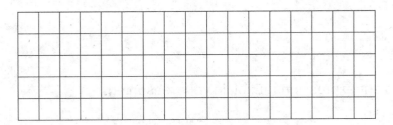

b. Draw a diagram of $x + 3$ and of $2x$ when x is 2.

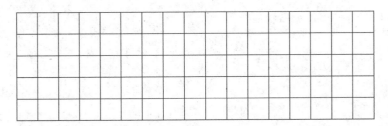

c. Draw a diagram of $x + 3$ and of $2x$ when x is 3.

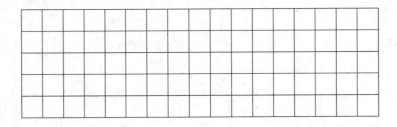

d. Draw a diagram of $x + 3$ and of $2x$ when x is 4.

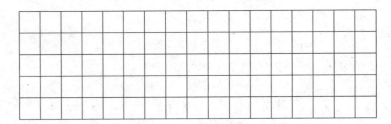

e. When are $x + 3$ and $2x$ equal? When are they not equal? Use your diagrams to explain.

2. Respond to each of the following.

   a. Do $4x$ and $15 + x$ have the same value when x is 5?

   b. Are $4x$ and $15 + x$ equivalent expressions? Explain your reasoning.

3. Respond to each of the following.

   a. Check that $2b + b$ and $3b$ have the same value when b is 1, 2, and 3.

   b. Do $2b + b$ and $3b$ have the same value for all values of b? Explain your reasoning.

   c. Are $2b + b$ and $3b$ equivalent expressions?

NAME _____ DATE _____ PERIOD _____

4. 80% of *x* is equal to 100. (Lesson 6-7)

   a. Write an equation that shows the relationship of 80%, *x*, and 100.

   b. Use your equation to find *x*.

5. For each story problem, write an equation to represent the problem and then solve the equation. Be sure to explain the meaning of any variables you use. (Lesson 6-5)

   a. Jada's dog was $5\frac{1}{2}$ inches tall when it was a puppy. Now her dog is $14\frac{1}{2}$ inches taller than that. How tall is Jada's dog now?

   b. Lin picked $9\frac{3}{4}$ pounds of apples, which was 3 times the weight of the apples Andre picked. How many pounds of apples did Andre pick?

6. Find these products. (Lesson 5-8)

   a. (2.3) · (1.4)

   b. (1.72) · (2.6)

   c. (18.2) · (0.2)

   d. 15 · (1.2)

7. Calculate 141.75 ÷ 2.5 using a method of your choice. Show or explain your reasoning. (Lesson 5-13)

Lesson 6-9

The Distributive Property, Part 1

NAME _____ DATE _____ PERIOD _____

Learning Goal Let's use the distributive property to make calculating easier.

Warm Up
9.1 Number Talk: Ways to Multiply

Find each product mentally.

1. $5 \cdot 102$ **2.** $5 \cdot 98$ **3.** $5 \cdot 999$

Activity
9.2 Ways to Represent Area of a Rectangle

1. Select **all** the expressions that represent the area of the large, outer rectangle in figure A. Explain your reasoning.

    $6 + 3 + 2$ $6 \cdot 5$

    $6 \cdot 3 + 6 \cdot 2$ $6 (3 + 2)$

    $6 \cdot 3 + 2$ $6 \cdot 3 \cdot 2$

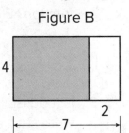

Figure A

Figure B

2. Select **all** the expressions that represent the area of the shaded rectangle on the left side of figure B. Explain your reasoning.

    $4 \cdot 7 + 4 \cdot 2$ $4 (7 - 2)$

    $4 \cdot 7 \cdot 2$ $4 (7 + 2)$

    $4 \cdot 5$ $4 \cdot 2 - 4 \cdot 7$

    $4 \cdot 7 - 4 \cdot 2$

Activity

9.3 Distributive Practice

Complete the table. If you get stuck, skip an entry and come back to it, or consider drawing a diagram of two rectangles that share a side.

Column 1	Column 2	Column 3	Column 4	Value
$5 \cdot 98$	$5(100 - 2)$	$5 \cdot 100 - 5 \cdot 2$	$500 - 10$	490
$33 \cdot 12$	$33(10 + 2)$			
		$3 \cdot 10 - 3 \cdot 4$	$30 - 12$	
	$100(0.04 + 0.06)$			
		$8 \cdot \frac{1}{2} + 8 \cdot \frac{1}{4}$		
			$9 + 12$	
			$24 - 16$	

Are you ready for more?

1. Use the distributive property to write two expressions that equal 360. (There are many correct ways to do this.)

2. Is it possible to write an expression like $a(b + c)$ that equals 360 where a is a fraction? Either write such an expression, or explain why it is impossible.

3. Is it possible to write an expression like $a(b - c)$ that equals 360? Either write such an expression, or explain why it is impossible.

4. How many ways do you think there are to make 360 using the distributive property?

NAME _____ DATE _____ PERIOD _____

Summary
The Distributive Property, Part 1

A **term** is a single number or variable, or variables and numbers multiplied together. Some examples of terms are:

- 10

- 8*x*

- *ab*

- 7*yz*

When we need to do mental calculations, we often come up with ways to make the calculation easier to do mentally.

Suppose we are grocery shopping and need to know how much it will cost to buy 5 cans of beans at 79 cents a can. We may calculate mentally in this way:

$5 \cdot 79$

$5 \cdot 70 + 5 \cdot 9$

$350 + 45$

395

In general, when we multiply two terms (or factors), we can break up one of the factors into parts, multiply each part by the other factor, and then add the products. The result will be the same as the product of the two original factors.

When we break up one of the factors and multiply the parts we are using the distributive property.

The distributive property also works with subtraction.

Here is another way to find $5 \cdot 79$:

$5 \cdot 79$

$5 \cdot (80 - 1)$

$400 - 5$

395

Glossary

term

Practice
The Distributive Property, Part 1

1. Select **all** the expressions that represent the area of the large, outer rectangle.

    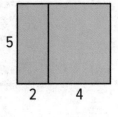

    (A.) $5(2 + 4)$

    (B.) $5 \cdot 2 + 4$

    (C.) $5 \cdot 2 + 5 \cdot 4$

    (D.) $5 \cdot 2 \cdot 4$

    (E.) $5 + 2 + 4$

    (F.) $5 \cdot 6$

2. Draw and label diagrams that show these two methods for calculating $19 \cdot 50$.

    a. First find $10 \cdot 50$ and then add $9 \cdot 50$.

    b. First find $20 \cdot 50$ and then take away 50.

3. Complete each calculation using the distributive property.

    a. $98 \cdot 24$
    $(100 - 2) \cdot 24$
    ...

    b. $21 \cdot 15$
    $(20 + 1) \cdot 15$
    ...

    c. $0.51 \cdot 40$
    $(0.5 + 0.01) \cdot 40$
    ...

NAME _____ DATE _____ PERIOD _____

4. A group of 8 friends goes to the movies. A bag of popcorn costs $2.99. How much will it cost to get one bag of popcorn for each friend? Explain how you can calculate this amount mentally.

5. Respond to each of the following.

   a. On graph paper, draw diagrams of $a + a + a + a$ and $4a$ when a is 1, 2, and 3. What do you notice? (Lesson 6-8)

   b. Do $a + a + a + a$ and $4a$ have the same value for any value of a? Explain how you know.

6. 120% of x is equal to 78. (Lesson 6-7)

    a. Write an equation that shows the relationship of 120%, x, and 78.

    b. Use your equation to find x. Show your reasoning.

7. Kiran's aunt is 17 years older than Kiran. (Lesson 6-6)

    a. How old will Kiran's aunt be when Kiran is:

        • 15 years old?

        • 30 years old?

        • x years old?

    b. How old will Kiran be when his aunt is 60 years old?

Lesson 6-10

The Distributive Property, Part 2

NAME _____ DATE _____ PERIOD _____

Learning Goal Let's use rectangles to understand the distributive property with variables.

 ## Warm Up
10.1 Possible Areas

1. A rectangle has a width of 4 units and a length of m units.
    Write an expression for the area of this rectangle.

2. What is the area of the rectangle if m is:

    • 3 units?

    • 2.2 units?

    • $\frac{1}{5}$ unit?

3. Could the area of this rectangle be 11 square units? Why or why not?

10.2 Partitioned Rectangles When Lengths Are Unknown

1. Here are two rectangles. The length and width of one rectangle are 8 and 5. The width of the other rectangle is 5, but its length is unknown so we labeled it *x*. Write an expression for the sum of the areas of the two rectangles.

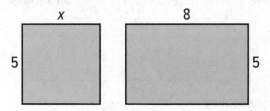

2. The two rectangles can be composed into one larger rectangle as shown. What are the width and length of the new, large rectangle?

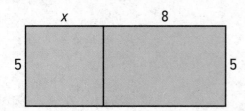

3. Write an expression for the total area of the large rectangle as the product of its width and its length.

NAME _____ DATE _____ PERIOD _____

 Activity

10.3 Areas of Partitioned Rectangles

For each rectangle, write expressions for the length and width and two expressions for the total area. Record them in the table. Check your expressions in each row with your group and discuss any disagreements.

A
a 5
3

B
6   x
$\frac{1}{3}$

C
1 1 1
r

D
$p\ p\ p\ p$
6

E
6   8
m

F
$3x$ 8
5

Rectangle	Width	Length	Area as a Product of Width Times Length	Area as a Sum of the Areas of the Smaller Rectangles
A				
B				
C				
D				
E				
F				

Here is an area diagram of a rectangle.

	y	z
w	A	24
x	18	72

1. Find the lengths w, x, y, and z, and the area A. All values are whole numbers.

2. Can you find another set of lengths that will work? How many possibilities are there?

Summary
The Distributive Property, Part 2

Here is a rectangle composed of two smaller rectangles A and B.

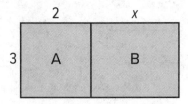

Based on the drawing, we can make several observations about the area of the rectangle:

- One side length of the large rectangle is 3 and the other is $2 + x$, so its area is $3(2 + x)$.

- Since the large rectangle can be decomposed into two smaller rectangles, A and B, with no overlap, the area of the large rectangle is also the sum of the areas of rectangles A and B: $3(2) + 3(x)$ or $6 + 3x$.

- Since both expressions represent the area of the large rectangle, they are equivalent to each other. $3(2 + x)$ is equivalent to $6 + 3x$.

We can see that multiplying 3 by the sum $2 + x$ is equivalent to multiplying 3 by 2 and then 3 by x and adding the two products. This relationship is an example of the *distributive property*.

$$3(2 + x) = 3 \cdot 2 + 3 \cdot x$$

NAME _____ DATE _____ PERIOD _____

Practice
The Distributive Property, Part 2

1. Here is a rectangle.

   a. Explain why the area of the large rectangle is $2a + 3a + 4a$.

   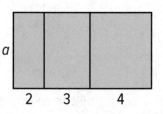

   b. Explain why the area of the large rectangle is $(2 + 3 + 4)a$.

2. Is the area of the shaded rectangle $6(2 - m)$ or $6(m - 2)$? Explain how you know.

   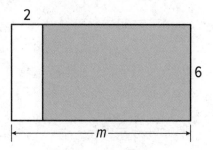

3. Choose the expressions that do *not* represent the total area of the rectangle. Select **all** that apply.

   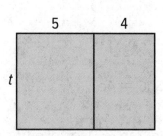

   A. $5t + 4t$

   B. $t + 5 + 4$

   C. $9t$

   D. $4 \cdot 5 \cdot t$

   E. $t(5 + 4)$

4. Evaluate each expression mentally. (Lesson 6-9)

a. $35 \cdot 91 - 35 \cdot 89$

b. $22 \cdot 87 + 22 \cdot 13$

c. $\frac{9}{11} \cdot \frac{7}{10} - \frac{9}{11} \cdot \frac{3}{10}$

5. Select **all** the expressions that are equivalent to $4b$. (Lesson 6-8)

(A.) $b + b + b + b$ (D.) $b \cdot b \cdot b \cdot b$

(B.) $b + 4$ (E.) $b \div \frac{1}{4}$

(C.) $2b + 2b$

6. Solve each equation. Show your reasoning. (Lesson 6-4)

a. $111 = 14a$ b. $13.65 = b + 4.88$

c. $c + \frac{1}{3} = 5\frac{1}{8}$ d. $\frac{2}{5}d = \frac{17}{4}$

e. $5.16 = 4e$

7. Andre ran $5\frac{1}{2}$ laps of a track in 8 minutes at a constant speed. It took Andre x minutes to run each lap. Select **all** the equations that represent this situation. (Lesson 6-2)

(A.) $\left(5\frac{1}{2}\right)x = 8$ (D.) $5\frac{1}{2} \div x = 8$

(B.) $5\frac{1}{2} + x = 8$ (E.) $x = 8 \div \left(5\frac{1}{2}\right)$

(C.) $5\frac{1}{2} - x = 8$ (F.) $x = \left(5\frac{1}{2}\right) \div 8$

Lesson 6-11

The Distributive Property, Part 3

NAME _____ DATE _____ PERIOD _____

Learning Goal Let's practice writing equivalent expressions by using the distributive property.

Warm Up
11.1 The Shaded Region

A rectangle with dimensions 6 cm and w cm is partitioned into two smaller rectangles.

Explain why each of these expressions represents the area, in cm², of the shaded region.

a. $6w - 24$

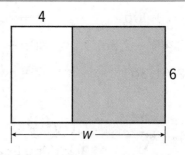

b. $6(w - 4)$

Activity

11.2 Matching to Practice Distributive Property

Match each expression in column 1 to an equivalent expression in column 2.
If you get stuck, consider drawing a diagram.

Column 1	Column 2
$a(1 + 2 + 3)$	$3(4a + b)$
$2(12 - 4)$	$12 \cdot 2 - 4 \cdot 2$
$12a + 3b$	$2(3a + 5b)$
$\frac{2}{3}(15a - 18)$	$(2 + 3)a$
$6a + 10b$	$a + 2a + 3a$
$0.4(5 - 2.5a)$	$10a - 12$
$2a + 3a$	$2 - a$

Activity

11.3 Writing Equivalent Expressions Using the Distributive Property

The distributive property can be used to write equivalent expressions.
In each row, use the distributive property to write an equivalent expression.
If you get stuck, consider drawing a diagram.

Product	Sum or Difference
$3(3 + x)$	
	$4x - 20$
$(9 - 5)x$	
	$4x + 7x$
$3(2x + 1)$	
	$10x - 5$
	$x + 2x + 3x$
$\frac{1}{2}(x - 6)$	
$y(3x + 4z)$	
	$2xyz - 3yz + 4xz$

NAME _____ DATE _____ PERIOD _____

Are you ready for more?

This rectangle has been cut up into squares of varying sizes. Both small squares have side length 1 unit. The square in the middle has side length x units.

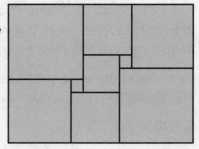

1. Suppose that x is 3. Find the area of each square in the diagram. Then find the area of the large rectangle.

2. Find the side lengths of the large rectangle assuming that x is 3. Find the area of the large rectangle by multiplying the length times the width. Check that this is the same area you found before.

3. Now suppose that we do not know the value of x. Write an expression for the side lengths of the large rectangle that involves x.

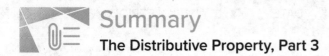

Summary
The Distributive Property, Part 3

The distributive property can be used to write a sum as a product, or write a product as a sum. You can always draw a partitioned rectangle to help reason about it, but with enough practice, you should be able to apply the distributive property without making a drawing.

Here are some examples of expressions that are equivalent due to the distributive property.

$9 + 18 = 9(1 + 2)$

$2(3x + 4) = 6x + 8$

$2n + 3n + n = n(2 + 3 + 1)$

$11b - 99a = 11(b - 9a)$

$k(c + d - e) = kc + kd - ke$

NAME _____ DATE _____ PERIOD _____

 Practice

The Distributive Property, Part 3

1. For each expression, use the distributive property to write an equivalent expression.

   a. $4(x + 2)$

   b. $(6 + 8) \cdot x$

   c. $4(2x + 3)$

   d. $6(x + y + z)$

2. Priya rewrites the expression $8y - 24$ as $8(y - 3)$. Han rewrites $8y - 24$ as $2(4y - 12)$. Are Priya's and Han's expressions each equivalent to $8y - 24$? Explain your reasoning.

3. Select **all** the expressions that are equivalent to $16x + 36$.

   (A.) $16(x + 20)$ (D.) $2(8x + 18)$

   (B.) $x(16 + 36)$ (E.) $2(8x + 36)$

   (C.) $4(4x + 9)$

4. The area of a rectangle is $30 + 12x$. List at least 3 possibilities for the length and width of the rectangle.

5. Select **all** the expressions that are equivalent to $\frac{1}{2}z$. (Lesson 6-8)

(A.) $z + z$

(B.) $z \div 2$

(C.) $z \cdot z$

(D.) $\frac{1}{4}z + \frac{1}{4}z$

(E.) $2z$

6. Respond to each of the following. (Lesson 6-6)

   a. What is the perimeter of a square with side length:

     3 cm   7 cm s cm

   b. If the perimeter of a square is 360 cm, what is its side length?

   c. What is the area of a square with side length:

     3 cm   7 cm s cm

   d. If the area of a square is 121 cm², what is its side length?

7. Solve each equation. (Lesson 6-5)

   a. $10 = 4a$ **b.** $5b = 17.5$

   c. $1.036 = 10c$ **d.** $0.6d = 1.8$

   e. $15 = 0.1e$

Lesson 6-12

Meaning of Exponents

NAME _____ DATE _____ PERIOD _____

Learning Goal Let's see how exponents show repeated multiplication.

 ## Warm Up
12.1 Notice and Wonder: Dots and Lines

What do you notice? What do you wonder?

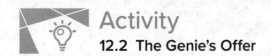

Activity

12.2 The Genie's Offer

You find a brass bottle that looks really old. When you rub some dirt off of the bottle, a genie appears! The genie offers you a reward. You must choose one:

- $50,000; or

- A magical $1 coin. The coin will turn into two coins on the first day. The two coins will turn into four coins on the second day. The four coins will double to 8 coins on the third day. The genie explains the doubling will continue for 28 days.

1. The number of coins on the third day will be $2 \cdot 2 \cdot 2$. Write an equivalent expression using exponents.

2. What do 2^5 and 2^6 represent in this situation? Evaluate 2^5 and 2^6 without a calculator. Pause for discussion.

3. How many days would it take for the number of magical coins to exceed $50,000?

4. Will the value of the magical coins exceed a million dollars within the 28 days? Explain or show your reasoning.

Are you ready for more?

A scientist is growing a colony of bacteria in a petri dish. She knows that the bacteria are growing and that the number of bacteria doubles every hour. When she leaves the lab at 5 p.m., there are 100 bacteria in the dish. When she comes back the next morning at 9 a.m., the dish is completely full of bacteria. At what time was the dish half full?

NAME _____ DATE _____ PERIOD _____

Activity

12.3 Make 81

1. Here are some expressions. All but one of them equals 16. Find the one that is *not* equal to 16 and explain how you know.

$2^3 \cdot 2$ $\qquad$ 4^2 $\qquad$ $\dfrac{2^5}{2}$ $\qquad$ 8^2

2. Write three expressions containing exponents so that each expression equals 81.

Summary

Meaning of Exponents

When we write an expression like 2^n, we call n the exponent.

If n is a positive whole number, it tells how many factors of 2 we should multiply to find the value of the expression.

For example:

- $2^1 = 2$, and

- $2^5 = 2 \cdot 2 \cdot 2 \cdot 2 \cdot 2$.

There are different ways to say 2^5. We can say

- "two raised to the power of five," or

- "two to the fifth power," or just

- "two to the fifth."

1. Select **all** expressions that are equivalent to 64.

   (A.) 2^6 (D.) 8^2

   (B.) 2^8 (E.) 16^4

   (C.) 4^3 (F.) 32^2

2. Select **all** the expressions that equal 3^4.

   (A.) 7   (D.) 81

   (B.) 4^3 (E.) 64

   (C.) 12   (F.) 9^2

3. 4^5 is equal to 1,024. Evaluate each expression.

   a. 4^6

   b. 4^4

   c. $4^3 \cdot 4^2$

NAME _____ DATE _____ PERIOD _____

4. $6^3 = 216$. Using exponents, write three more expressions whose value is 216.

5. Find two different ways to rewrite $3xy + 6yz$ using the distributive property. **(Lesson 6-11)**

6. Solve each equation. **(Lesson 6-5)**

   a. $a - 2.01 = 5.5$

   b. $b + 2.01 = 5.5$

   c. $10c = 13.71$

   d. $100d = 13.71$

7. Which expressions represent the total area of the large rectangle? Select **all** that apply. **(Lesson 6-10)**

   $\textbf{A.}$ $6(m + n)$

   $\textbf{B.}$ $6n + m$

   $\textbf{C.}$ $6n + 6m$

   $\textbf{D.}$ $6mn$

   $\textbf{E.}$ $(n + m)6$

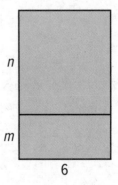

8. Is each statement true or false? Explain your reasoning. (Lesson 3-16)

a. $\dfrac{45}{100} \cdot 72 = \dfrac{45}{72} \cdot 100$

b. 16% of 250 is equal to 250% of 16

Lesson 6-13

Expressions with Exponents

NAME _____ DATE _____ PERIOD _____

Learning Goal Let's use the meaning of exponents to decide if equations are true.

Warm Up
13.1 Which One Doesn't Belong: Twos

Which one doesn't belong?

$2 \cdot 2 \cdot 2 \cdot 2$ 16 2^4 $4 \cdot 2$

Activity
13.2 Is the Equation True?

Decide whether each equation is true or false, and explain how you know.

1. $2^4 = 2 \cdot 4$

2. $3 + 3 + 3 + 3 + 3 = 3^5$

3. $5^3 = 5 \cdot 5 \cdot 5$

4. $2^3 = 3^2$

5. $16^1 = 8^2$

6. $\frac{1}{2} \cdot \frac{1}{2} \cdot \frac{1}{2} \cdot \frac{1}{2} = 4 \cdot \frac{1}{2}$

7. $\left(\frac{1}{2}\right)^4 = \frac{1}{8}$

8. $8^2 = 4^3$

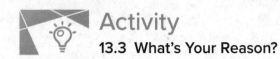
In each list, find expressions that are equivalent to each other and explain to your partner why they are equivalent. Your partner listens to your explanation. If you disagree, explain your reasoning until you agree. Switch roles for each list.

(There may be more than two equivalent expressions in each list.)

1. $5 \cdot 5$

   2^5

   5^2

   $2 \cdot 5$

2. 4^3

   3^4

   $4 \cdot 4 \cdot 4$

   $4 + 4 + 4$

3. $6 + 6 + 6$

   6^3

   3^6

   $3 \cdot 6$

4. 11^5

   $11 \cdot 11 \cdot 11 \cdot 11 \cdot 11$

   $11 \cdot 5$

   5^{11}

5. $\dfrac{1}{5} \cdot \dfrac{1}{5} \cdot \dfrac{1}{5}$

   $\left(\dfrac{1}{5}\right)^3$

   $\dfrac{1}{15}$

   $\dfrac{1}{125}$

6. $\left(\dfrac{5}{3}\right)^2$

   $\left(\dfrac{3}{5}\right)^2$

   $\dfrac{10}{6}$

   $\dfrac{25}{9}$

NAME _____ DATE _____ PERIOD _____

Are you ready for more?

What is the last digit of $3^{1,000}$? Show or explain your reasoning.

Summary
Expressions with Exponents

When working with exponents, the bases don't have to always be whole numbers. They can also be other kinds of numbers, like fractions, decimals, and even variables.

For example, we can use exponents in each of the following ways:

$$\left(\frac{2}{3}\right)^4 = \frac{2}{3} \cdot \frac{2}{3} \cdot \frac{2}{3} \cdot \frac{2}{3}$$

$$(1.7)^3 = (1.7) \cdot (1.7) \cdot (1.7)$$

$$x^5 = x \cdot x \cdot x \cdot x \cdot x$$

Practice
Expressions with Exponents

1. Select **all** the expressions that are equal to $3 \cdot 3 \cdot 3 \cdot 3 \cdot 3$.

   (A.) $3 \cdot 5$

   (B.) 3^5

   (C.) $3^4 \cdot 3$

   (D.) $5 \cdot 3$

   (E.) 5^3

2. Noah starts with 0 and then adds the number 5 four times. Diego starts with 1 and then multiplies by the number 5 four times. For each expression, decide whether it is equal to Noah's result, Diego's result, or neither.

   a. $4 \cdot 5$

   b. $4 + 5$

   c. 4^5

   d. 5^4

3. Decide whether each equation is true or false, and explain how you know.

   a. $9 \cdot 9 \cdot 3 = 3^5$

   b. $7 + 7 + 7 = 3 + 3 + 3 + 3 + 3 + 3 + 3$

   c. $\frac{1}{7} \cdot \frac{1}{7} \cdot \frac{1}{7} = \frac{3}{7}$

   d. $4^1 = 4 \cdot 1$

   e. $6 + 6 + 6 = 6^3$

NAME _____ DATE _____ PERIOD _____

4. Respond to each of the following.

   a. What is the area of a square with side lengths of $\frac{3}{5}$ units?

   b. What is the side length of a square with area $\frac{1}{16}$ square units?

   c. What is the volume of a cube with edge lengths of $\frac{2}{3}$ units?

   d. What is the edge length of a cube with volume $\frac{27}{64}$ cubic units?

5. Select **all** the expressions that represent the area of the shaded rectangle. (Lesson 6-10)

   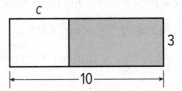

   (A.) $3(10 - c)$ (D.) $10(3 - c)$

   (B.) $3(c - 10)$ (E.) $30 - 3c$

   (C.) $10(c - 3)$ (F.) $30 - 10c$

6. A ticket at a movie theater costs $8.50. One night, the theater had $29,886 in ticket sales. (Lesson 5-13)

   a. Estimate about how many tickets the theater sold. Explain your reasoning.

   b. How many tickets did the theater sell? Explain your reasoning.

7. A fence is being built around a rectangular garden that is $8\frac{1}{2}$ feet by $6\frac{1}{3}$ feet. Fencing comes in panels. Each panel is $\frac{2}{3}$ of a foot wide. How many panels are needed? Explain or show your reasoning. (Lesson 4-12)

Lesson 6-14

Evaluating Expressions with Exponents

NAME _____ DATE _____ PERIOD _____

Learning Goal Let's find the values of expressions with exponents.

Warm Up
14.1 Revisiting the Cube

Based on the given information, what other measurements of the square and cube could we find?

Activity
14.2 Calculating Surface Area

A cube has side length 10 inches.

- Jada says the surface area of the cube is 600 in^2.

- Noah says the surface area of the cube is 3,600 in^2.

Here is how each of them reasoned:

Jada's Method:

$6 \cdot 10^2$

$6 \cdot 100$

600

Noah's Method:

$6 \cdot 10^2$

60^2

$3,600$

Do you agree with either of them? Explain your reasoning.

Activity
14.3 Expression Explosion

Evaluate the expressions in one of the columns. Your partner will work on the other column. Check with your partner after you finish each row. Your answers in each row should be the same. If your answers aren't the same, work together to find the error.

Column A	Column B
$5^2 + 4$	$2^2 + 25$
$2^4 \cdot 5$	$2^3 \cdot 10$
$3 \cdot 4^2$	$12 \cdot 2^2$
$20 + 2^3$	$1 + 3^3$
$9 \cdot 2^1$	$3 \cdot 6^1$
$\frac{1}{9} \cdot \left(\frac{1}{2}\right)^3$	$\frac{1}{8} \cdot \left(\frac{1}{3}\right)^2$

Are you ready for more?

1. Consider this equation: $\boxed{}^2 + \boxed{}^2 = \boxed{}^2$. An example of 3 different whole numbers that could go in the boxes are 3, 4, and 5, since $3^2 + 4^2 = 5^2$. (That is, $9 + 16 = 25$) Can you find a different set of 3 different whole numbers that make the equation true?

2. How many sets of 3 different whole numbers can you find?

3. Can you find a set of 3 different whole numbers that make this equation true?

$$\boxed{}^3 + \boxed{}^3 = \boxed{}^3$$

4. How about this one?

$$\boxed{}^4 + \boxed{}^4 = \boxed{}^4$$

5. Once you have worked on this a little while, you can understand a problem that is famous in the history of math. (Alas, this space is too small to contain it.) If you are interested, consider doing some further research on *Fermat's Last Theorem*.

NAME _____ DATE _____ PERIOD _____

Summary
Evaluating Expressions with Exponents

Exponents give us a new way to describe operations with numbers, so we need to understand how exponents get along with the other operations we know.

When we write $6 \cdot 4^2$, we want to make sure everyone agrees about how to evaluate this. Otherwise some people might multiply first and others compute the exponent first, and different people would get different values for the same expression!

Earlier we saw situations in which $6 \cdot 4^2$ represented the surface area of a cube with side lengths 4 units. When computing the surface area, we evaluate 4^2 first (or find the area of one face of the cube first) and then multiply the result by 6. In many other expressions that use exponents, the part with an exponent is intended to be evaluated first.

To make everyone agree about the value of expressions like $6 \cdot 4^2$, the convention is to *evaluate the part of the expression with the exponent first.*

Here are a couple of examples:

$6 \cdot 4^2$ $45 + 5^2$

$6 \cdot 16$ $45 + 25$

96 70

If we want to communicate that 6 and 4 should be multiplied first and then squared, then we can use parentheses to group parts together:

$(6 \cdot 4)^2$ $(45 + 5)^2$

24^2 50^2

576 $2{,}500$

Practice
Evaluating Expressions with Exponents

1. Lin says, "I took the number 8, and then multiplied it by the square of 3."
   Select **all** the expressions that equal Lin's answer.

   (A.) $8 \cdot 3^2$

   (B.) $(8 \cdot 3)^2$

   (C.) $8 \cdot 2^3$

   (D.) $3^2 \cdot 8$

   (E.) 24^2

   (F.) 72

2. Evaluate each expression.

   a. $7 + 2^3$ b. $9 \cdot 3^1$

   c. $20 - 2^4$ d. $2 \cdot 6^2$

   e. $8 \cdot \left(\frac{1}{2}\right)^2$ f. $\frac{1}{3} \cdot 3^3$

   g. $\left(\frac{1}{5} \cdot 5\right)^5$

NAME _____ DATE _____ PERIOD _____

3. Andre says, "I multiplied 4 by 5, then cubed the result."
   Select **all** the expressions that equal Andre's answer.

   (A.) $4 \cdot 5^3$

   (B.) $(4 \cdot 5)^3$

   (C.) $(4 \cdot 5)^2$

   (D.) $5^3 \cdot 4$

   (E.) 20^3

   (F.) 500

   (G.) 8,000

4. Han has 10 cubes, each 5 inches on a side.

   a. Find the total volume of Han's cubes. Express your answer as an
      expression using an exponent.

   b. Find the total surface area of Han's cubes. Express your answer as an
      expression using an exponent.

5. Priya says that $\frac{1}{3} \cdot \frac{1}{3} \cdot \frac{1}{3} \cdot \frac{1}{3} = \frac{4}{3}$. Do you agree with Priya? Explain or show
   your reasoning. (Lesson 6-13)

6. Answer each question. Show your reasoning. (Lesson 6-7)

   a. 125% of e is 30. What is e?

   b. 35% of f is 14. What is f?

7. Which expressions are solutions to the equation $2.4y = 13.75$?
   Select **all** that apply. (Lesson 6-5)

   (A.) $13.75 - 1.4$

   (B.) $13.75 \cdot 2.4$

   (C.) $13.75 \div 2.4$

   (D.) $\frac{13.75}{2.4}$

   (E.) $2.4 \div 13.75$

8. Jada explains how she finds $15 \cdot 23$:

   "I know that ten 23s is 230, so five 23s will be half of 230, which is 115.
   15 is 10 plus 5, so $15 \cdot 23$ is 230 plus 115, which is 345." (Lesson 5-7)

   a. Do you agree with Jada? Explain.

   b. Draw a 15 by 23 rectangle. Partition the rectangle into two rectangles
      and label them to show Jada's reasoning.

Lesson 6-15

Equivalent Exponential Expressions

NAME _____ DATE _____ PERIOD _____

Learning Goal Let's investigate expressions with variables and exponents.

 ## Warm Up
15.1 Up or Down?

1. Respond to each of the following.

   a. Find the values of 3^x and $\left(\frac{1}{3}\right)^x$ for different values of x.

x	3^x	$\left(\frac{1}{3}\right)^x$
1		
2		
3		
4		

   b. What patterns do you notice?

 ## Activity
15.2 What's the Value?

Evaluate each expression for the given value of x.

1. $3x^2$ when x is 10

2. $3x^2$ when x is $\frac{1}{9}$

3. $\frac{x^3}{4}$ when x is 4

4. $\frac{x^3}{4}$ when x is $\frac{1}{2}$

5. $9 + x^7$ when x is 1

6. $9 + x^7$ when x is $\frac{1}{2}$

Activity

15.3 Exponent Experimentation

Find a solution to each equation in the list. (Numbers in the list may be a solution to more than one equation, and not all numbers in the list will be used.)

List: $\dfrac{8}{125}$ $\dfrac{6}{15}$ $\dfrac{5}{8}$ $\dfrac{8}{9}$ 1 $\dfrac{4}{3}$ 2 3 4 5 6 8

1. $64 = x^2$

2. $64 = x^3$

3. $2^x = 32$

4. $x = \left(\dfrac{2}{5}\right)^3$

5. $\dfrac{16}{9} = x^2$

6. $2 \cdot 2^5 = 2^x$

7. $2x = 2^4$

8. $4^3 = 8^x$

NAME _____ DATE _____ PERIOD _____

Are you ready for more?

This fractal is called a Sierpinski Tetrahedron. A tetrahedron is a polyhedron that has four faces. (The plural of tetrahedron is tetrahedra.) The small tetrahedra form four medium-sized tetrahedra: blue, red, yellow, and green. The medium-sized tetrahedra form one large tetrahedron.

1. How many small faces does this fractal have? Be sure to include faces you can't see as well as those you can. Try to find a way to figure this out so that you don't have to count every face.

2. How many small tetrahedra are in the bottom layer, touching the table?

3. To make an even bigger version of this fractal, you could take four fractals like the one pictured and put them together. Explain where you would attach the fractals to make a bigger tetrahedron.

4. How many small faces would this bigger fractal have? How many small tetrahedra would be in the bottom layer?

5. What other patterns can you find?

Summary
Equivalent Exponential Expressions

In this lesson, we saw expressions that used the letter x as a variable. We evaluated these expressions for different values of x.

- To evaluate the expression $2x^3$ when x is 5, we replace the letter x with 5 to get $2 \cdot 5^3$. This is equal to $2 \cdot 125$ or just 250. So the value of $2x^3$ is 250 when x is 5.

- To evaluate $\frac{x^2}{8}$ when x is 4, we replace the letter x with 4 to get $\frac{4^2}{8} = \frac{16}{8}$, which equals 2. So $\frac{x^2}{8}$ has a value of 2 when x is 4.

We also saw equations with the variable x and had to decide what value of x would make the equation true.

- Suppose we have an equation $10 \cdot 3^x = 90$ and a list of possible solutions:

  1, 2, 3, 9, 11.

  The only value of x that makes the equation true is 2 because $10 \cdot 3^2 = 10 \cdot 3 \cdot 3$, which equals 90. So 2 is the solution to the equation.

NAME _____ DATE _____ PERIOD _____

 Practice

Equivalent Exponential Expressions

1. Evaluate each expression if $x = 3$.

   a. 2^x

   b. x^2

   c. 1^x

   d. x^1

   e. $\left(\dfrac{1}{2}\right)^x$

2. Evaluate each expression for the given value of each variable.

   a. $2 + x^3$, x is 3

   b. x^2, x is $\dfrac{1}{2}$

   c. $3x^2 + y$, x is 5, y is 3

   d. $10y + x^2$, x is 6, y is 4

3. Decide if the expressions have the same value. If not, determine which expression has the larger value.

   a. 2^3 and 3^2

   b. 1^{31} and 31^1

   c. 4^2 and 2^4

   d. $\left(\dfrac{1}{2}\right)^3$ and $\left(\dfrac{1}{3}\right)^2$

4. Match each equation to its solution.

Equations	Solutions
a. $7 + x^2 = 16$	$x = 1$
b. $5 - x^2 = 1$	$x = 2$
c. $2 \cdot 2^3 = 2^x$	$x = 3$
d. $\dfrac{3^4}{3^x} = 27$	$x = 4$

5. An adult pass at the amusement park costs 1.6 times as much as a child's pass. (Lesson 6-6)

   a. How many dollars does an adult pass cost if a child's pass costs:

      • $5?

      • $10?

      • *w* dollars?

   b. A child's pass costs $15. How many dollars does an adult pass cost?

6. Jada reads 5 pages every 20 minutes. At this rate, how many pages can she read in 1 hour? (Lesson 2-14)

   a. Use a double number line to find the answer.

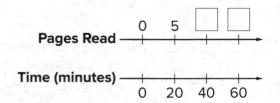

   b. Use a table to find the answer.

Pages Read	Time (minutes)
5	20

   c. Which strategy do you think is better, and why?

Lesson 6-16

Two Related Quantities, Part 1

NAME _____ DATE _____ PERIOD _____

Learning Goal Let's use equations and graphs to describe relationships with ratios.

Warm Up
16.1 Which One Would You Choose?

Which one would you choose? Be prepared to explain your reasoning.

• A 5-pound jug of honey for $15.35

• Three 1.5-pound jars of honey for $13.05

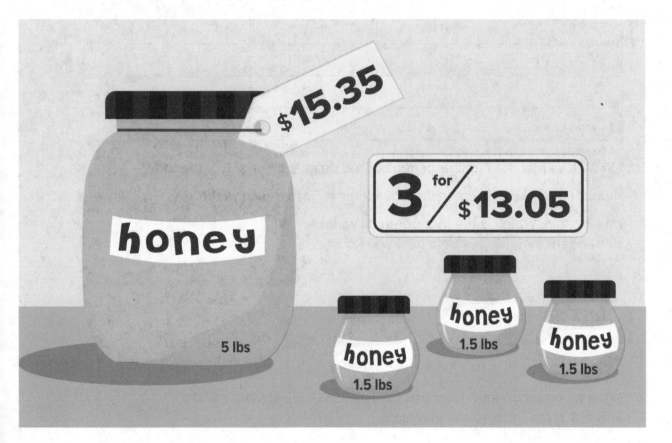

Activity

16.2 Painting the Set

Lin needs to mix a specific shade of orange paint for the set of the school play. The color uses 3 parts yellow for every 2 parts red.

1. Complete the table to show different combinations of red and yellow paint that will make the shade of orange Lin needs.

Cups of Red Paint (r)	Cups of Yellow Paint (y)	Total Cups of Paint (t)
2	3	
6		
		20
	18	
14		
16		
		50
	42	

2. Lin notices that the number of cups of red paint is always $\frac{2}{5}$ of the total number of cups. She writes the equation $r = \frac{2}{5}t$ to describe the relationship. Which is the **independent variable**? Which is the **dependent variable**? Explain how you know.

3. Write an equation that describes the relationship between r and y where y is the independent variable.

4. Write an equation that describes the relationship between y and r where r is the independent variable.

NAME _____ DATE _____ PERIOD _____

5. Use the points in the table to create two graphs that show the relationship between *r* and *y*. Match each relationship to one of the equations you wrote.

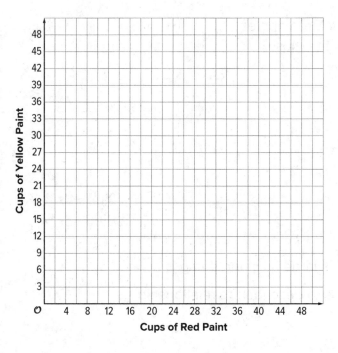

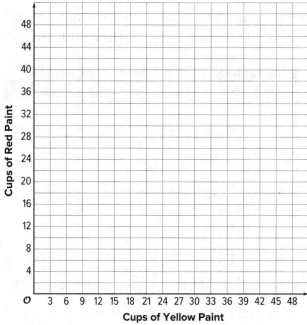

Are you ready for more?

A fruit stand sells apples, peaches, and tomatoes. Today, they sold 4 apples for every 5 peaches. They sold 2 peaches for every 3 tomatoes. They sold 132 pieces of fruit in total. How many of each fruit did they sell?

Equations are very useful for describing sets of equivalent ratios.

Here is an example.

A pie recipe calls for 3 green apples for every 5 red apples.
We can create a table to show some equivalent ratios.

Green Apples (g)	Red Apples (r)
3	5
6	10
9	15
12	20

We can see from the table that r is always $\frac{5}{3}$ as large as g and that g is always $\frac{3}{5}$ as large as r. We can write equations to describe the relationship between g and r.

- When we know the number of green apples and want to find the number of red apples, we can write: $r = \frac{5}{3}g$. In this equation, if g changes, r is affected by the change, so we refer to g as the **independent variable** and r as the **dependent variable**.

  We can use this equation with any value of g to find r. If 270 green apples are used, then $\frac{5}{3} \cdot (270)$ or 450 red apples are used.

- When we know the number of red apples and want to find the number of green apples, we can write: $g = \frac{3}{5}r$. In this equation, if r changes, g is affected by the change, so we refer to r as the independent variable and g as the dependent variable.

  We can use this equation with any value of r to find g. If 275 red apples are used, then $\frac{3}{5} \cdot (275)$ or 165 green apples are used.

NAME _____ DATE _____ PERIOD _____

We can also graph the two equations we wrote to get a visual picture of the relationship between the two quantities.

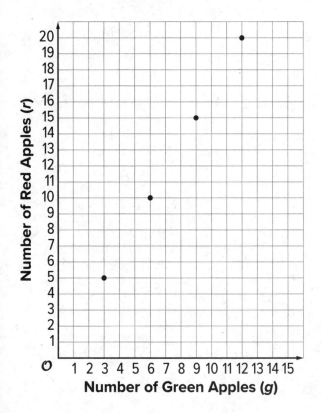
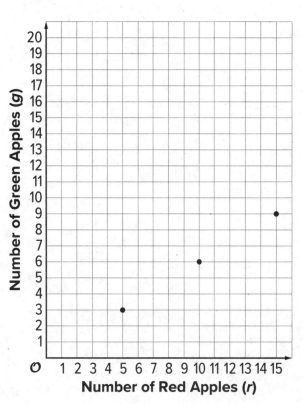

Glossary

dependent variable

independent variable

Practice

Two Related Quantities, Part 1

1. Here is a graph that shows some values for the number of cups of sugar, *s*, required to make *x* batches of brownies.

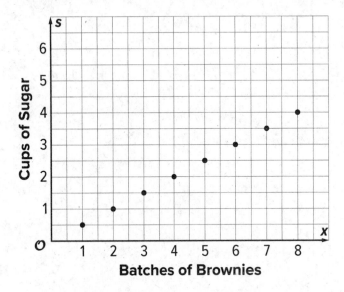

Cups of Sugar

Batches of Brownies

a. Complete the table so that the pair of numbers in each column represents the coordinates of a point on the graph.

x	1	2	3	4	5	6	7
s							

b. What does the point (8, 4) mean in terms of the amount of sugar and number of batches of brownies?

c. Write an equation that shows the amount of sugar in terms of the number of batches.

NAME _____ DATE _____ PERIOD _____

2. Each serving of a certain fruit snack contains 90 calories.

   a. Han wants to know how many calories he gets from the fruit snacks. Write an equation that shows the number of calories, c, in terms of the number of servings, n.

   b. Tyler needs some extra calories each day during his sports season. He wants to know how many servings he can have each day if all the extra calories come from the fruit snack. Write an equation that shows the number of servings, n, in terms of the number of calories, c.

3. Kiran shops for books during a 20% off sale.

   a. What percent of the original price of a book does Kiran pay during the sale?

   b. Complete the table to show how much Kiran pays for books during the sale.

   c. Write an equation that relates the sale price, s, to the original price, p.

   d. On graph paper, create a graph showing the relationship between the sale price and the original price by plotting the points from the table.

Original Price in dollars (p)	Sale Price in dollars (s)
1	
2	
3	
4	
5	
6	
7	
8	
9	
10	

4. Evaluate each expression when x is 4 and y is 6. **(Lesson 6-15)**

    a. $(6 - x)^3 + y$ **b.** $2 + x^3$

    c. $2^x - 2y$ **d.** $\left(\dfrac{1}{2}\right)^x$

    e. $1^x + 2^x$ **f.** $\dfrac{2^x}{x^2}$

5. Find $(12.34) \cdot (0.7)$. Show your reasoning. **(Lesson 5-8)**

6. For each expression, write another division expression that has the same value and that can be used to help find the quotient. Then, find each quotient. **(Lesson 5-13)**

    a. $302.1 \div 0.5$

    b. $12.15 \div 0.02$

    c. $1.375 \div 0.11$

Lesson 6-17

Two Related Quantities, Part 2

NAME _____ DATE _____ PERIOD _____

Learning Goal Let's use equations and graphs to describe stories with constant speed.

Warm Up
17.1 Walking to the Library

Lin and Jada each walk at a steady rate from school to the library. Lin can walk 13 miles in 5 hours, and Jada can walk 25 miles in 10 hours. They each leave school at 3:00 and walk $3\frac{1}{4}$ miles to the library. What time do they each arrive?

Activity
17.2 The Walk-a-thon

Diego, Elena, and Andre participated in a walk-a-thon to raise money for cancer research. They each walked at a constant rate, but their rates were different.

1. Complete the table to show how far each participant walked during the walk-a-thon.

Time (hours)	Miles Walked by Diego	Miles Walked by Elena	Miles Walked by Andre
1			
2	6		
	12	11	
5			17.5

2. How fast was each participant walking in miles per hour?

3. How long did it take each participant to walk one mile?

4. Graph the progress of each person in the **coordinate plane.** Use a different color for each participant.

5. Diego says that $d = 3t$ represents his walk, where d is the distance walked in miles and t is the time in hours.

    a. Explain why $d = 3t$ relates the distance Diego walked to the time it took.

    b. Write two equations that relate distance and time: one for Elena and one for Andre.

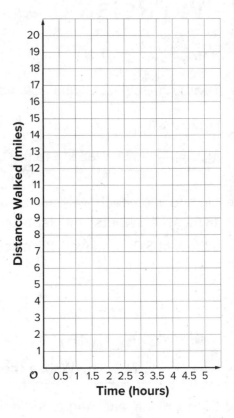

6. Use the equations you wrote to predict how far each participant would walk, at their same rate, in 8 hours.

7. For Diego's equation and the equations you wrote, which is the dependent variable and which is the independent variable?

NAME _____ DATE _____ PERIOD _____

Are you ready for more?

1. Two trains are traveling toward each other, on parallel tracks. Train A is moving at a constant speed of 70 miles per hour. Train B is moving at a constant speed of 50 miles per hour. The trains are initially 320 miles apart. How long will it take them to meet? One way to start thinking about this problem is to make a table. Add as many rows as you like.

	Train A	Train B
Starting Position	0 miles	320 miles
After 1 hour	70 miles	270 miles
After 2 hours		

2. How long will it take a train traveling at 120 miles per hour to go 320 miles?

3. Explain the connection between these two problems.

Equations are very useful for solving problems with constant speeds. Here is an example.

A boat is traveling at a constant speed of 25 miles per hour.

- How far can the boat travel in 3.25 hours?

- How long does it take for the boat to travel 60 miles?

We can write equations to help us answer questions like these. Let's use t to represent the time in hours and d to represent the distance in miles that the boat travels.

When we know the time and want to find the distance, we can write: $d = 25t$	When we know the distance and want to find the time, we can write: $t = \dfrac{d}{25}$
In this equation, if t changes, d is affected by the change, so t is the independent variable and d is the dependent variable.	In this equation, if d changes, t is affected by the change, so d is the independent variable and t is the dependent variable.
This equation can help us find d when we have any value of t. In 3.25 hours, the boat can travel 25 (3.25) or 81.25 miles.	This equation can help us find t when for any value of d. To travel 60 miles, it will take $\dfrac{60}{25}$ or $2\dfrac{2}{5}$ hours.

These problems can also be solved using important ratio techniques such as a table of equivalent ratios. The equations are particularly valuable in this case because the answers are not round numbers or easy to quickly evaluate.

We can also graph the two equations we wrote to get a visual picture of the relationship between the two quantities:

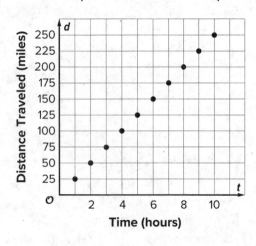
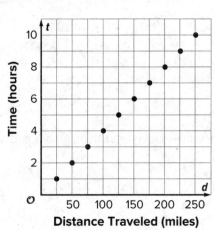

Glossary

coordinate plane

NAME _____ DATE _____ PERIOD _____

Practice
Two Related Quantities, Part 2

1. A car is traveling down a road at a constant speed of 50 miles per hour.

   a. Complete the table with the amounts of time it takes the car to travel certain distances, or the distances traveled for certain amounts of time.

Time (hours)	Distance (miles)
2	
1.5	
t	
	50
	300
	d

   b. Write an equation that represents the distance traveled by the car, d, for an amount of time, t.

   c. In your equation, which is the dependent variable and which is the independent variable?

2. The graph represents the amount of time in hours it takes a ship to travel various distances in miles.

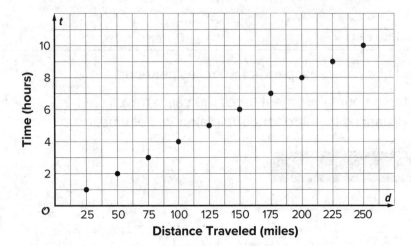

a. Write the coordinates of one point on the graph. What does the point represent?

b. What is the speed of the ship in miles per hour?

c. Write an equation that relates the time, t, it takes to travel a given distance, d.

NAME _____ DATE _____ PERIOD _____

3. Find a solution to each equation in the list that follows.
   (Not all numbers will be used.) **(Lesson 6-15)**

   List: $\frac{1}{10}$ $\frac{1}{3}$ 1   2   3   4   5   7   8   10   16

   a. $2^x = 8$ b. $2^x = 2$

   c. $x^2 = 100$ d. $x^2 = \frac{1}{100}$

   e. $x^1 = 7$ f. $2^x \cdot 2^3 = 2^7$

   g. $\frac{2^x}{2^3} = 2^5$

4. Select **all** the expressions that are equivalent to $5x + 30x - 15x$. **(Lesson 6-11)**

   (A.) $5(x + 6x - 3x)$

   (B.) $(5 + 30 - 15) \cdot x$

   (C.) $x(5 + 30x - 15x)$

   (D.) $5x(1 + 6 - 3)$

   (E.) $5(x + 30x - 15x)$

5. Evaluate each expression if x is 1, y is 2, and z is 3. (Lesson 6-15)

a. $7x^2 - z$

b. $(x + 4)^3 - y$

c. $y(x + 3^3)$

d. $(7 - y + z)^2$

e. $0.241x + x^3$

Lesson 6-18

More Relationships

NAME _____ DATE _____ PERIOD _____

Learning Goal Let's use graphs and equations to show relationships involving area, volume, and exponents.

Warm Up
18.1 Which One Doesn't Belong: Graphs

Which one doesn't belong?

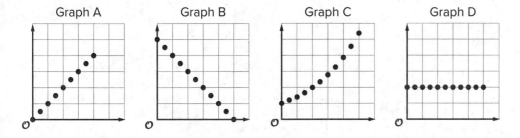

Graph A   Graph B   Graph C   Graph D

Activity
18.2 Making a Banner

Mai is creating a rectangular banner to advertise the school play. The material for the banner is sold by the square foot. Mai has enough money to buy 36 square feet of material. She is trying to decide on the length and width of the banner.

1. If the length is 6 feet, what is the width?

2. If the length is 4 feet, what is the width?

3. If the length is 9 feet, what is the width?

4. To find different combinations of length and width that give an area of 36 square feet, Mai uses the equation $w = \frac{36}{\ell}$, where w is the width and ℓ is the length. Compare your strategy and Mai's method for finding the width. How were they the same or different?

5. Use several combinations of length and width to create a graph that shows the relationship between the side lengths of various rectangles with area 36 square feet.

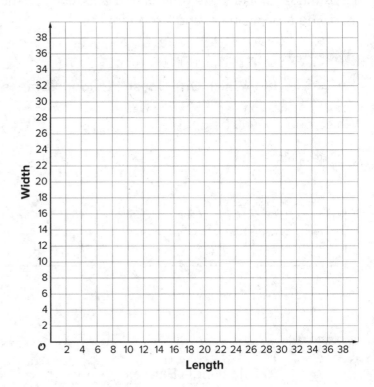

6. Explain how the graph describes the relationship between length and width for different rectangles with area 36.

7. Suppose Mai used the equation $\ell = \frac{36}{w}$ to find the length for different values of the width. Would the graph be different if she graphed length on the vertical axis and width on the horizontal axis? Explain how you know.

NAME _____ DATE _____ PERIOD _____

Activity

18.3 Cereal Boxes

A cereal manufacturer needs to design a cereal box that has a volume of 225 cubic inches and a height that is no more than 15 inches.

1. The designers know that the volume of a rectangular prism can be calculated by multiplying the area of its base and its height. Complete the table with pairs of values that will make the volume 225 in³.

Height (in)		5	9	12		$7\frac{1}{2}$
Area of Base (in²)	75				15	

2. Describe how you found the missing values for the table.

3. Write an equation that shows how the area of the base, A, is affected by changes in the height, h, for different rectangular prisms with volume 225 in³.

4. Plot the ordered pairs from the table on the graph to show the relationship between the area of the base and the height for different boxes with volume 225 in³.

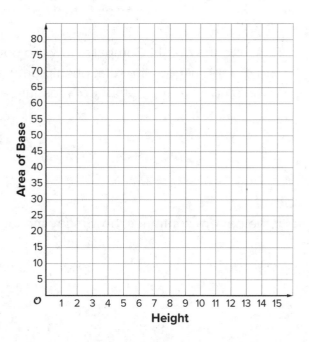

Activity

18.4 Multiplying Mosquitoes

A researcher who is studying mosquito populations collects the following data:

Day in the Study (d)	Number of Mosquitoes (n)
1	2
2	4
3	8
4	16
5	32

1. The researcher said that, for these five days, the number of mosquitoes, n, can be found with the equation $n = 2^d$ where d is the day in the study. Explain why this equation matches the data.

2. Use the ordered pairs in the table to graph the relationship between number of mosquitoes and day in the study for these five days.

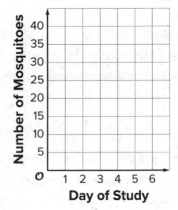

3. Describe the graph. Compare how the data, equation, and graph illustrate the relationship between the day in the study and the number of mosquitoes.

4. If the pattern continues, how many mosquitoes will there be on Day 6?

NAME _____ DATE _____ PERIOD _____

Summary
More Relationships

Equations can represent relationships between geometric quantities. For instance:

- If s is the side length of a square, then the area A is related to s.

$$A = s^2$$

- Sometimes the relationships are more specific. For example, the perimeter P of a rectangle with length l and width w is $P = 2l + 2w$. If we consider only rectangles with a length of 10, then the relationship between the perimeter and the width is.

$$P = 20 + 2w$$

Here is another example of an equation with exponent expressing the relationship between quantities.

- A super ball is dropped from 10 feet. On each successive bounce, it only goes $\frac{1}{2}$ as high as on the previous bounce.

  This means that on the first bounce, the ball will bounce 5 feet high, and then on the second bounce it will only go $2\frac{1}{2}$ feet high, and so on. We can represent this situation with an equation to find how high the super ball will bounce after any number of bounces.

  To find how high the super ball bounces on the nth bounce, we have to multiply 10 feet (the intial height) by $\frac{1}{2}$ and multiply by $\frac{1}{2}$ again for each bounce thereafter; we need to do this n times. So the height, h, of the ball on the nth bounce will be:

$$h = 10\left(\frac{1}{2}\right)^n$$

  In this equation, the dependent variable, h, is affected by changes in the independent variable, n.

Equations and graphs can give us insight into different kinds of relationships between quantities and help us answer questions and solve problems.

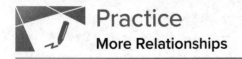

Practice
More Relationships

1. Elena is designing a logo in the shape of a parallelogram. She wants the logo to have an area of 12 square inches. She draws bases of different lengths and tries to compute the height for each.

   a. Write an equation Elena can use to find the height, h, for each value of the base, b.

   b. Use your equation to find the height of a parallelogram with base 1.5 inches.

2. Han is planning to ride his bike 24 miles.

   a. How long will it take if he rides at a rate of:

   3 miles per hour?

   4 miles per hour?

   6 miles per hour?

   b. Write an equation that Han can use to find t, the time it will take to ride 24 miles, if his rate in miles per hour is represented by r.

NAME _____ DATE _____ PERIOD _____

   c. On graph paper, draw a graph that shows t in terms of r for a 24-mile ride.

3. The graph of the equation $V = 10s^3$ contains the points (2, 80) and (4, 640).

   a. Create a story that is represented by this graph.

   b. What do the points mean in the context of your story?

4. You find a brass bottle that looks really old. When you rub some dirt off the bottle, a genie appears! The genie offers you a reward. You must choose one:

- $50,000; or

- A magical $1 coin. The coin will turn into two coins on the first day. The two coins will turn into four coins on the second day. The four coins will double to 8 coins on the third day. The genie explains the doubling will continue for 28 days.

   a. Write an equation that shows the number of coins, n, in terms of the day, d.

   b. Create a table that shows the number of coins for each day for the first 15 days.

NAME _____ DATE _____ PERIOD _____

c. Create a graph for days 7 through 12 that shows how the number of coins grows with each day.

5. At a market, 3.1 pounds of peaches cost $7.72. How much did the peaches cost per pound? Explain or show your reasoning. Round your answer to the nearest cent. **(Lesson 5-13)**

6. Andre set up a lemonade stand last weekend. It cost him $0.15 to make each cup of lemonade, and he sold each cup for $0.35. **(Lesson 5-13)**

   a. If Andre collects $9.80, how many cups did he sell?

   b. How much money did it cost Andre to make this amount of lemonade?

   c. How much money did Andre make in profit?

Lesson 6-19

Tables, Equations, and Graphs, Oh My!

NAME _____ DATE _____ PERIOD _____

Learning Goal Let's explore some equations from real-world situations.

 Activity

19.1 Matching Equations and Tables

Match each equation with a table that represents the same relationship.
Be prepared to explain your reasoning.

$S - 2 = T$ $G = J + 13$ $P = I - 47.50$ $C + 273.15 = K$ $e = 6s$

$m = 8.96V$ $y = \frac{1}{12}x$ $t = \frac{d}{2.5}$ $g = 28.35z$

Table 1:

Independent Variable	Dependent Variable
20	8
58.85	23.54
804	321.6

Table 2:

Independent Variable	Dependent Variable
5	18
36	49
75	88

Table 3:

Independent Variable	Dependent Variable
2.5	22.4
20	179.2
75	672

Table 4:

Independent Variable	Dependent Variable
20	$1\frac{2}{3}$
36	3
804	67

Table 5:

Independent Variable	Dependent Variable
58.85	11.35
175.5	128
804	756.5

Table 6:

Independent Variable	Dependent Variable
2.5	275.65
20	293.15
58.85	332

Table 7:

Independent Variable	Dependent Variable
5	3
20	18
36	34

Table 8:

Independent Variable	Dependent Variable
2.6	73.71
20	567
36	1,020.6

Table 9:

Independent Variable	Dependent Variable
2.6	15.6
36	216
58.85	353.1

Activity

19.2 Getting to Know an Equation

The equations in the previous activity represent situations.

- $S - 2 = T$ where S is the number of sides on a polygon and T is the number of triangles you can draw inside it (from one vertex to the others, without overlapping)

- $G = J + 13$ where G is a day in the Gregorian calendar and J is the same day in the Julian calendar

- $P = I - 47.50$ where I is the amount of income and P is the profit after $47.50 in expenses

- $C + 273.15 = K$ where C is a temperature in degrees Celsius and K is the same temperature in Kelvin

- $e = 6s$ where e is the total edge length of a tetrahedron and s is the length of one side

- $m = 8.96V$ where V is the volume of a piece of copper and m is its mass

- $y = \frac{1}{12}x$ where x is the number of eggs and y is how many dozens that makes

- $t = \frac{d}{2.5}$ where t is the amount of time it takes in seconds to jog a distance of d meters at a constant speed of 2.5 meters per second

- $g = 28.35z$ where g is the mass in grams and z is the same amount in ounces

Your teacher will assign you one of these equations to examine more closely.

1. Rewrite your equation using words. Use words like product, sum, difference, quotient, and term.

NAME _____ DATE _____ PERIOD _____

2. In the previous activity, you matched equations and tables. Copy the values from the table that matched your assigned equation into the first 3 rows of this table. Make sure to label what each column represents.

Independent Variable: _____	Dependent Variable: _____
60	
	300

3. Select one of the first 3 rows of the table and explain what those values mean in this situation.

4. Use your equation to find the values that complete the last 2 rows of the table. Explain your reasoning.

5. On graph paper, create a graph that represents this relationship. Make sure to label your axes.

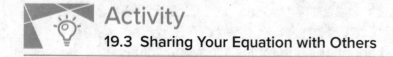

Activity

19.3 Sharing Your Equation with Others

Create a visual display of your assigned relationships that includes:

• your equation along with an explanation of each variable

• a verbal description of the relationship

• your table

• your graph

If you have time, research more about your relationship and add more details or illustrations to help explain the situation.

Learning Targets

Lesson	Learning Target(s)
6-1 Tape Diagrams and Equations	• I can tell whether or not an equation could represent a tape diagram.
	• I can use a tape diagram to represent a situation.
6-2 Truth and Equations	• I can match equations to real life situations they could represent.
	• I can replace a variable in an equation with a number that makes the equation true, and know that this number is called a solution to the equation.
6-3 Staying in Balance	• I can compare doing the same thing to the weights on each side of a balanced hanger to solving equations by subtracting the same amount from each side or dividing each side by the same number.
	• I can explain what a balanced hanger and a true equation have in common.
	• I can write equations that could represent the weights on a balanced hanger.

(continued on the next page)

(continued from the previous page)

Lesson	Learning Target(s)
6-4 Practice Solving Equations and Representing Situations with Equations	• I can explain why different equations can describe the same situation. • I can solve equations that have whole numbers, fractions, and decimals.
6-5 A New Way to Interpret *a* over *b*	• I understand the meaning of a fraction made up of fractions or decimals, like $\frac{2.1}{0.07}$ or $\frac{\frac{4}{5}}{\frac{3}{2}}$. • When I see an equation, I can make up a story that the equation might represent, explain what the variable represents in the story, and solve the equation.
6-6 Write Expressions Where Letters Stand for Numbers	• I can use an expression that represents a situation to find an amount in a story. • I can write an expression with a variable to represent a calculation where I do not know one of the numbers.
6-7 Revisit Percentages	• I can solve percent problems by writing and solving an equation.

Lesson	Learning Target(s)
6-8 Equal and Equivalent	• I can explain what it means for two expressions to be equivalent.
	• I can use a tape diagram to figure out when two expressions are equal.
	• I can use what I know about operations to decide whether two expressions are equivalent.
6-9 The Distributive Property, Part 1	• I can use a diagram of a rectangle split into two smaller rectangles to write different expressions representing its area.
	• I can use the distributive property to help do computations in my head.
6-10 The Distributive Property, Part 2	• I can use a diagram of a split rectangle to write different expressions with variables representing its area.
6-11 The Distributive Property, Part 3	• I can use the distributive property to write equivalent expressions with variables.

(continued on the next page)

(continued from the previous page)

Lesson	Learning Target(s)
6-12 Meaning of Exponents	• I can evaluate expressions with exponents and write expressions with exponents that are equal to a given number. • I understand the meaning of an expression with an exponent like 3^5.
6-13 Expressions with Exponents	• I can decide if expressions with exponents are equal by evaluating the expressions or by understanding what exponents mean.
6-14 Evaluating Expressions with Exponents	• I know how to evaluate expressions that have both an exponent and addition or subtraction. • I know how to evaluate expressions that have both an exponent and multiplication or division.
6-15 Equivalent Exponential Expressions	• I can find solutions to equations with exponents in a list of numbers. • I can replace a variable with a number in an expression with exponents and operations and use the correct order to evaluate the expression.

Lesson	Learning Target(s)
6-16 Two Related Quantities, Part 1	• I can create tables and graphs that show the relationship between two amounts in a given ratio. • I can write an equation with variables that shows the relationship between two amounts in a given ratio.
6-17 Two Related Quantities, Part 2	• I can create tables and graphs to represent the relationship between distance and time for something moving at a constant speed. • I can write an equation with variables to represent the relationship between distance and time for something moving at a constant speed.
6-18 More Relationships	• I can create tables and graphs that show different kinds of relationships between amounts. • I can write equations that describe relationships with area and volume.
6-19 Tables, Equations, and Graphs, Oh My!	• I can create a table and a graph that represent the relationship in a given equation. • I can explain what an equation tells us about the situation.

(continued on the next page)

Notes:

Unit 7

Rational Numbers

How can you describe the location of fish and other animals that live below the ocean's surface? In this unit, you'll see how rational numbers can help you.

Topics

- Negative Numbers and Absolute Value
- Inequalities
- The Coordinate Plane
- Common Factors and Common Multiples
- Let's Put It to Work

Unit 7

Rational Numbers

Lesson 7-1

Positive and Negative Numbers

NAME _____ DATE _____ PERIOD _____

Learning Goal Let's explore how we represent temperatures and elevations.

Warm Up

1.1 Notice and Wonder: Memphis and Bangor

What do you notice? What do you wonder?

Activity

1.2 Above and Below Zero

1. Here are three situations involving changes in temperature and three number lines. Represent each change on a number line. Then, answer the question.

   a. At noon, the temperature was 5 degrees Celsius. By late afternoon, it has risen 6 degrees Celsius. What was the temperature late in the afternoon?

   b. The temperature was 8 degrees Celsius at midnight. By dawn, it has dropped 12 degrees Celsius. What was the temperature at dawn?

   c. Water freezes at 0 degrees Celsius, but the freezing temperature can be lowered by adding salt to the water. A student discovered that adding half a cup of salt to a gallon of water lowers its freezing temperature by 7 degrees Celsius. What is the freezing temperature of the gallon of salt water?

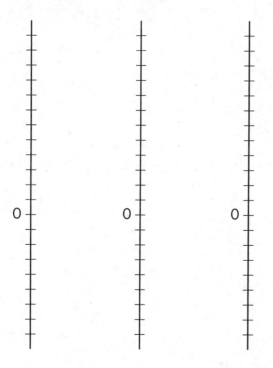

2. Discuss with a partner:

   a. How did each of you name the resulting temperature in each situation?

   b. What does it mean when the temperature is above 0? Below 0?

   c. Do numbers less than 0 make sense in other contexts? Give some specific examples to show how they do or do not make sense.

NAME _____ DATE _____ PERIOD _____

Activity
1.3 High Places, Low Places

1. Here is a table that shows elevations of various cities.

City	Elevation (feet)
Harrisburg, PA	320
Bethell, IN	1,211
Denver, CO	5,280
Coachella, CA	-22
Death Valley, CA	-282
New York City, NY	33
Miami, FL	0

a. On the list of cities, which city has the second highest elevation?

b. How would you describe the elevation of Coachella, CA in relation to sea level?

c. How would you describe the elevation of Death Valley, CA in relation to sea level?

d. If you are standing on a beach right next to the ocean, what is your elevation?

e. How would you describe the elevation of Miami, FL?

f. A city has a higher elevation than Coachella, CA. Select all numbers that could represent the city's elevation. Be prepared to explain your reasoning.

-11 feet   -35 feet   4 feet   -8 feet   0 feet

2. Here are two tables that show the elevations of highest points on land and lowest points in the ocean. Distances are measured from sea level.

Mountain	Continent	Elevation (meters)
Everest	Asia	8,848
Kilimanjaro	Africa	5,895
Denali	North America	6,168
Pikchu Pikchu	South America	5,664

Trench	Ocean	Elevation (meters)
Mariana Trench	Pacific	-11,033
Puerto Rico Trench	Atlantic	-8,600
Tonga Trench	Pacific	-10,882
Sunda Trench	Indian	-7,725

a. Which point in the ocean is the lowest in the world? What is its elevation?

b. Which mountain is the highest in the world? What is its elevation?

c. If you plot the elevations of the mountains and trenches on a vertical number line, what would 0 represent? What would points above 0 represent? What about points below 0?

d. Which is farther from sea level: the deepest point in the ocean, or the top of the highest mountain in the world? Explain.

NAME _____ DATE _____ PERIOD _____

Are you ready for more?

A spider spins a web in the following way:

- It starts at sea level.
- It moves up one inch in the first minute.
- It moves down two inches in the second minute.
- It moves up three inches in the third minute.
- It moves down four inches in the fourth minute.

Assuming that the pattern continues, what will the spider's elevation be after an hour has passed?

Positive numbers are numbers that are greater than 0.
Negative numbers are numbers that are less than zero.
The meaning of a negative number in a context depends
on the meaning of zero in that context.

For example, if we measure temperatures in degrees Celsius,
then 0 degrees Celsius corresponds to the temperature at
which water freezes.

In this context, positive temperatures are warmer than the
freezing point and negative temperatures are colder than
the freezing point. A temperature of -6 degrees Celsius
means that it is 6 degrees away from 0 and it is less than 0.

This thermometer shows a temperature of -6 degrees Celsius.

If the temperature rises a few degrees and gets very close to 0 degrees
without reaching it, the temperature is still a negative number.

Another example is elevation, which is a distance above or below
sea level. An elevation of 0 refers to the sea level. Positive
elevations are higher than sea level, and negative elevations
are lower than sea level.

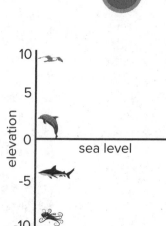

Glossary

negative number
positive number

NAME _____ DATE _____ PERIOD _____

Practice
Positive and Negative Numbers

1. Respond to each of the following.

   a. Is a temperature of -11 degrees warmer or colder than a temperature of -15 degrees?

   b. Is an elevation of -10 feet closer or farther from the surface of the ocean than an elevation of -8 feet?

   c. It was 8 degrees at nightfall. The temperature dropped 10 degrees by midnight. What was the temperature at midnight?

   d. A diver is 25 feet below sea level. After he swims up 15 feet toward the surface, what is his elevation?

2. Respond to each of the following.

   a. A whale is at the surface of the ocean to breathe. What is the whale's elevation?

   b. The whale swims down 300 feet to feed. What is the whale's elevation now?

   c. The whale swims down 150 more feet more. What is the whale's elevation now?

   d. Plot each of the three elevations as a point on a vertical number line. Label each point with its numeric value.

3. Explain how to calculate a number that is equal to $\frac{2.1}{1.5}$. (Lesson 6-5)

4. Write an equation to represent each situation and then solve the equation. (Lesson 6-4)

    a. Andre drinks 15 ounces of water, which is $\frac{3}{5}$ of a bottle. How much does the bottle hold? Use x for the number of ounces of water the bottle holds.

    b. A bottle holds 15 ounces of water. Jada drank 8.5 ounces of water. How many ounces of water are left in the bottle? Use y for the number of ounces of water left in the bottle.

    c. A bottle holds z ounces of water. A second bottle holds 16 ounces, which is $\frac{8}{5}$ times as much water. How much does the first bottle hold?

5. A rectangle has an area of 24 square units and a side length of $2\frac{3}{4}$ units. Find the other side length of the rectangle. Show your reasoning. (Lesson 4-13)

Lesson 7-2

Points on the Number Line

NAME _____ DATE _____ PERIOD _____

Learning Goal Let's plot positive and negative numbers on the number line.

Warm Up
2.1 A Point on the Number Line

Which of the following numbers could be *B*?

2.5 $\dfrac{2}{5}$ $\dfrac{5}{2}$ $\dfrac{25}{10}$ 2.49

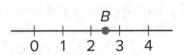

Activity
2.2 What's the Temperature?

1. Here are five thermometers. The first four thermometers show temperatures in Celsius. Write the temperatures in the blanks. The last thermometer is missing some numbers. Write them in the boxes.

a. _____ b. _____ c. _____ d. _____ e.

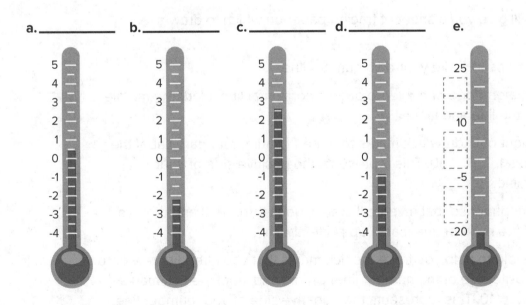

2. Elena says that the thermometer shown here reads -2.5°C because the line of the liquid is above -2°C. Jada says that it is -1.5°C. Do you agree with either one of them? Explain your reasoning.

3. One morning, the temperature in Phoenix, Arizona, was 8°C and the temperature in Portland, Maine, was 12°C cooler. What was the temperature in Portland?

Activity

2.3 Folded Number Lines

Your teacher will give you a sheet of tracing paper on which to draw a number line.

1. Follow the steps to make your own number line.

- Use a straightedge or a ruler to draw a horizontal line. Mark the middle point of the line and label it 0.

- To the right of 0, draw tick marks that are 1 centimeter apart. Label the tick marks 1, 2, 3. . . 10. This represents the positive side of your number line.

- Fold your paper so that a vertical crease goes through 0 and the two sides of the number line match up perfectly.

- Use the fold to help you trace the tick marks that you already drew onto the opposite side of the number line. Unfold and label the tick marks -1, -2, -3 . . . -10. This represents the negative side of your number line.

NAME _____ DATE _____ PERIOD _____

2. Use your number line to answer these questions:

   a. Which number is the same distance away from zero as is the number 4?

   b. Which number is the same distance away from zero as is the number -7?

   c. Two numbers that are the same distance from zero on the number line are called **opposites**. Find another pair of opposites on the number line.

   d. Determine how far away the number 5 is from 0. Then, choose a positive number and a negative number that are each farther away from zero than is the number 5.

   e. Determine how far away the number -2 is from 0. Then, choose a positive number and a negative number that is each farther away from zero than is the number -2.

   Pause here so your teacher can review your work.

3. Here is a number line with some points labeled with letters. Determine the location of points *P*, *X*, and *Y*.

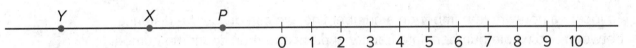

   If you get stuck, trace the number line and points onto a sheet of tracing paper, fold it so that a vertical crease goes through 0, and use the folded number line to help you find the unknown values.

   Are you ready for more?

   At noon, the temperatures in Portland, Maine, and Phoenix, Arizona, had opposite values. The temperature in Portland was 18°C lower than in Phoenix. What was the temperature in each city? Explain your reasoning.

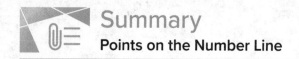
Here is a number line labeled with positive and negative numbers. The number 4 is positive, so its location is 4 units to the right of 0 on the number line. The number -1.1 is negative, so its location is 1.1 units to the left of 0 on the number line.

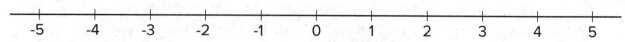

We say that the *opposite* of 8.3 is -8.3, and that the *opposite* of $\frac{-3}{2}$ is $\frac{3}{2}$. Any pair of numbers that are equally far from 0 are called **opposites**.

Points A and B are opposites because they are both 2.5 units away from 0, even though A is to the left of 0 and B is to the right of 0.

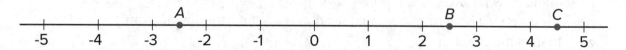

A positive number has a negative number for its opposite. A negative number has a positive number for its opposite. The opposite of 0 is itself.

You have worked with positive numbers for many years. All of the positive numbers you have seen—whole and non-whole numbers—can be thought of as fractions and can be located on a number line.

To locate a non-whole number on a number line, we can divide the distance between two whole numbers into fractional parts and then count the number of parts. For example, 2.7 can be written as $2\frac{7}{10}$. The segment between 2 and 3 can be partitioned into 10 equal parts or 10 tenths. From 2, we can count 7 of the tenths to locate 2.7 on the number line.

All of the fractions and their opposites are what we call **rational numbers**.

For example, 4, -1.1, 8.3, -8.3, $\frac{-3}{2}$, and $\frac{3}{2}$ are all rational numbers.

Glossary

opposite
rational number

NAME _____ DATE _____ PERIOD _____

Practice
Points on the Number Line

1. For each number, name its opposite.

   a. -5

   b. 28

   c. -10.4

   d. 0.875

   e. 0

   f. -8,003

2. Plot the numbers -1.5, $\frac{3}{2}$, -$\frac{3}{2}$, and -$\frac{4}{3}$ on the number line. Label each point with its numeric value.

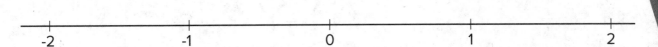

3. Plot the following points on a number line.

   a. -1.5

   b. the opposite of -2

   c. the opposite of 0.5

   d. -2

4. Respond to each of the following. (Lesson 7-1)

   a. Represent each of these temperatures in degrees Fahrenheit with a positive or negative number.

   i. 5 degrees above zero

   ii. 3 degrees below zero

   iii. 6 degrees above zero

   iv. $2\frac{3}{4}$ degrees below zero

   b. Order the temperatures above from the coldest to the warmest.

5. Solve each equation. (Lesson 6-5)

   a. $8x = \frac{2}{3}$

   b. $1\frac{1}{2} = 2x$

   c. $5x = \frac{2}{7}$

   d. $\frac{1}{4}x = 5$

   e. $\frac{1}{5} = \frac{2}{3}x$

6. Write the solution to each equation as a fraction and as a decimal. (Lesson 6-5)

   a. $2x = 3$

   b. $5y = 3$

   c. $0.3z = 0.009$

7. There are 15.24 centimeters in 6 inches. (Lesson 3-4)

   a. How many centimeters are in 1 foot?

   b. How many centimeters are in 1 yard?

Lesson 7-3

Comparing Positive and Negative Numbers

NAME _____ DATE _____ PERIOD _____

Learning Goal Let's compare numbers on the number line.

Warm Up
3.1 Which One Doesn't Belong: Inequalities

Which inequality doesn't belong?

$\frac{5}{4} < 2$ $8.5 > 0.95$

$8.5 < 7$ $10.00 < 100$

Activity
3.2 Comparing Temperatures

Here are the low temperatures, in degrees Celsius, for a week in Anchorage, Alaska.

Day	Mon	Tues	Weds	Thurs	Fri	Sat	Sun
Temperature	5	-1	-5.5	-2	3	4	0

1. Plot the temperatures on a number line. Which day of the week had the lowest low temperature?

2. The lowest temperature ever recorded in the United States was -62 degrees Celsius, in Prospect Creek Camp, Alaska. The average temperature on Mars is about -55 degrees Celsius.

   a. Which is warmer, the coldest temperature recorded in the USA, or the average temperature on Mars? Explain how you know.

   b. Write an inequality to show your answer.

3. On a winter day the low temperature in Anchorage, Alaska, was -21 degrees Celsius and the low temperature in Minneapolis, Minnesota, was -14 degrees Celsius. Jada said, "I know that 14 is less than 21, so -14 is also less than -21. This means that it was colder in Minneapolis than in Anchorage."

Do you agree? Explain your reasoning.

Are you ready for more?

Another temperature scale frequently used in science is the *Kelvin scale*. In this scale, 0 is the lowest possible temperature of anything in the universe, and it is -273.15 degrees in the Celsius scale. Each 1 K is the same as 1°C, so 10 K is the same as -263.15°C.

1. Water boils at 100°C. What is this temperature in K?

2. Ammonia boils at -35.5°C. What is the boiling point of ammonia in K?

3. Explain why only positive numbers (and 0) are needed to record temperature in K.

Activity

3.3 Rational Numbers on a Number Line

1. Plot the numbers -2, 4, -7, and 10 on the number line. Label each point with its numeric value.

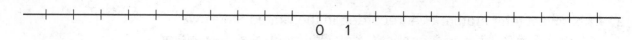

2. Decide whether each inequality statement is true or false. Be prepared to explain your reasoning.

   a. -2 < 4

   b. -2 < -7

   c. 4 > -7

   d. -7 > 10

3. Andre says that $\frac{1}{4}$ is less than $-\frac{3}{4}$ because, of the two numbers, $\frac{1}{4}$ is closer to 0. Do you agree? Explain your reasoning.

NAME _____ DATE _____ PERIOD _____

4. Answer each question. Be prepared to explain how you know.

   a. Which number is greater: $\frac{1}{4}$ or $\frac{5}{4}$?

   b. Which is farther from 0: $\frac{1}{4}$ or $\frac{5}{4}$?

   c. Which number is greater: $-\frac{3}{4}$ or $\frac{5}{8}$?

   d. Which is farther from 0: $-\frac{3}{4}$ or $\frac{5}{8}$?

   e. Is the number that is farther from 0 always the greater number? Explain your reasoning.

Summary
Comparing Positive and Negative Numbers

We use the words *greater than* and *less than* to compare numbers on the number line. For example, the numbers -2.7, 0.8, and -1.3, are shown on the number line.

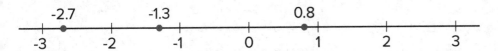

Because -2.7 is to the left of -1.3, we say that -2.7 is less than -1.3. We write: -2.7 < -1.3. In general, any number that is to the left of a number *n* is less than *n*.

We can see that -1.3 is greater than -2.7 because -1.3 is to the right of -2.7. We write: -1.3 > -2.7. In general, any number that is to the right of a number *n* is greater than *n*.

We can also see that 0.8 > -1.3 and 0.8 > -2.7. In general, any positive number is greater than any negative number.

> **Glossary**
>
> **sign**

Practice

Comparing Positive and Negative Numbers

1. Decide whether each inequality statement is true or false. Explain your reasoning.

   a. -5 > 2

   b. 3 > -8

   c. -12 > -15

   d. -12.5 > -12

2. Here is a true statement: -8.7 < -8.4. Select **all** of the statements that are equivalent to -8.7 < -8.4.

   (A.) -8.7 is further to the right on the number line than -8.4.

   (B.) -8.7 is further to the left on the number line than -8.4.

   (C.) -8.7 is less than -8.4.

   (D.) -8.7 is greater than -8.4.

   (E.) -8.4 is less than -8.7.

   (F.) -8.4 is greater than -8.7.

NAME _____ DATE _____ PERIOD _____

3. Plot each of the following numbers on the number line. Label each point with its numeric value. (Lesson 7-2)

0.4, -1.5, $-1\frac{7}{10}$, $-\frac{11}{10}$

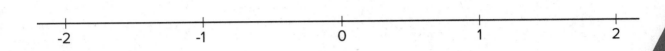

4. The table shows five states and the lowest point in each state. Put the states in order by their lowest elevation, from least to greatest. (Lesson 7-4)

State	Lowest Elevation (feet)
California	-282
Colorado	3350
Louisiana	-8
New Mexico	2842
Wyoming	3099

5. Each lap around the track is 400 meters. (Lesson 6-6)

    a. How many meters does someone run if they run:

       2 laps?

       5 laps?

       x laps?

    b. If Noah ran 14 laps, how many meters did he run?

    c. If Noah ran 7,600 meters, how many laps did he run?

6. A stadium can seat 16,000 people at full capacity. (Lesson 3-16)

    a. If there are 13,920 people in the stadium, what percentage of the capacity is filled? Explain or show your reasoning.

    b. What percentage of the capacity is not filled?

Lesson 7-4

Ordering Rational Numbers

NAME _____ DATE _____ PERIOD _____

Learning Goal Let's order rational numbers.

Warm Up
4.1 How Do They Compare?

1. Use the symbols >, <, or = to compare each pair of numbers. Be prepared to explain your reasoning.

   a. 12 _____ 19

   b. 212 _____ 190

   c. 15 _____ 1.5

   d. 9.02 _____ 9.2

   e. 6.050 _____ 6.05

   f. 0.4 _____ $\dfrac{9}{40}$

   g. $\dfrac{19}{24}$ _____ $\dfrac{19}{21}$

   h. $\dfrac{16}{17}$ _____ $\dfrac{11}{12}$

Activity
4.2 Ordering Rational Number Cards

Your teacher will give you a set of number cards. Order them from least to greatest.

Your teacher will give you a second set of number cards. Add these to the correct places in the ordered set.

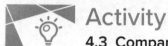

Activity

4.3 Comparing Points on a Line

1. Use each of the following terms at least once to describe or compare the values of points *M, N, P, R.*

   • greater than

   • less than

   • opposite of (or opposites)

   • negative number

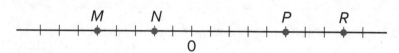

2. Tell what the value of each point would be if:

   a. *P* is $2\frac{1}{2}$

   b. *N* is -0.4

   c. *R* is 200

   d. *M* is -15

NAME _____ DATE _____ PERIOD _____

Are you ready for more?

The list of fractions between 0 and 1 with denominators between 1 and 3 looks like this: $\frac{0}{1}, \frac{1}{1}, \frac{1}{2}, \frac{1}{3}, \frac{2}{3}$. We can put them in order like this: $\frac{0}{1} < \frac{1}{3} < \frac{1}{2} < \frac{2}{3} < \frac{1}{1}$.

Now let's expand the list to include fractions with denominators of 4. We won't include $\frac{2}{4}$, because $\frac{1}{2}$ is already on the list: $\frac{0}{1} < \frac{1}{4} < \frac{1}{3} < \frac{1}{2} < \frac{2}{3} < \frac{3}{4} < \frac{1}{1}$.

1. Expand the list again to include fractions that have denominators of 5.

2. Expand the list you made to include fractions that have denominators of 6.

3. When you add a new fraction to the list, you put it in between two "neighbors." Go back and look at your work. Do you see a relationship between a new fraction and its two neighbors?

Summary
Ordering Rational Numbers

To order rational numbers from least to greatest, we list them in the order they appear on the number line from left to right.

For example, we can see that the numbers -2.7, -1.3, 0.8 are listed from least to greatest because of the order they appear on the number line.

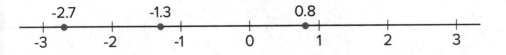

Practice

Ordering Rational Numbers

1. Select **all** of the numbers that are *greater than* -5.

   (A.) 1.3

   (B.) -6

   (C.) -12

   (D.) $\frac{1}{7}$

   (E.) -1

   (F.) -4

2. Order these numbers from least to greatest: $\frac{1}{2}$, 0, 1, $-1\frac{1}{2}$, $-\frac{1}{2}$, -1

3. Here are the boiling points of certain elements in degrees Celsius:

   • Argon: -185.8

   • Chlorine: -34

   • Fluorine: -188.1

   • Hydrogen: -252.87

   • Krypton: -153.2

   List the elements from least to greatest boiling points.

NAME _____ DATE _____ PERIOD _____

4. Explain why zero is considered its own opposite. **(Lesson 7- 2)**

5. Explain how to make these calculations mentally. **(Lesson 6-9)**

    a. $99 + 54$

    b. $244 - 99$

    c. $99 \cdot 6$

    d. $99 \cdot 15$

6. Find the quotients. (Lesson 4-11)

   a. $\frac{1}{2} \div 2$

   b. $2 \div 2$

   c. $\frac{1}{2} \div \frac{1}{2}$

   d. $\frac{38}{79} \div \frac{38}{79}$

7. Over several months, the weight of a baby measured in pounds doubles. Does its weight measured in kilograms also double? Explain. (Lesson 3-4)

Lesson 7-5

Using Negative Numbers to Make Sense of Contexts

NAME _____ DATE _____ PERIOD _____

Learning Goal Let's make sense of negative amounts of money.

 ## Warm Up
5.1 Notice and Wonder: It Comes and Goes

What do you notice? What do you wonder?

Activity	Amount
Do my chores	30.00
Babysit my cousin	45.00
Buy my lunch	-10.80
Get my allowance	15.00
Buy a shirt	-18.69
Pet my dog	0.00

Activity

5.2 The Concession Stand

The manager of the concession stand keeps records of all of the supplies she buys and all of the items she sells. The table shows some of her records for Tuesday.

Item	Quantity	Value in dollars
Doughnuts	-58	37.70
Straws	3,000	-10.35
Hot Dogs	-39	48.75
Pizza	13	-116.87
Apples	-40	14.00
French Fries	-88	132.00

1. Which items did she sell? Explain your reasoning.

2. How can we interpret -58 in this situation?

3. How can we interpret -10.35 in this situation?

4. On which item did she spend the most amount of money? Explain your reasoning.

NAME _____ DATE _____ PERIOD _____

Activity
5.3 Drinks for Sale

A vending machine in an office building sells bottled beverages. The machine keeps track of all changes in the number of bottles from sales and from machine refills and maintenance. This record shows the changes for every 5-minute period over one hour.

1. What might a positive number mean in this context? What about a negative number?

2. What would a "0" in the second column mean in this context?

3. Which numbers—positive or negative—result in fewer bottles in the machine?

4. At what time was there the greatest change to the number of bottles in the machine? How did that change affect the number of remaining bottles in the machine?

Time	Number of Bottles
8:00–8:04	-1
8:05–8:09	+12
8:10–8:14	-4
8:15–8:19	-1
8:20–8:24	-5
8:25–8:29	-12
8:30–8:34	-2
8:35–8:39	0
8:40–8:44	0
8:45–8:49	-6
8:50–8:54	+24
8:55–8:59	0.
Service	

5. At which time period, 8:05–8:09 or 8:25–8:29, was there a greater change to the number of bottles in the machine? Explain your reasoning.

6. The machine must be emptied to be serviced. If there are 40 bottles in the machine when it is to be serviced, what number will go in the second column in the table?

Priya, Mai, and Lin went to a cafe on a weekend. Their shared bill came to $25. Each student gave the server a $10 bill. The server took this $30 and brought back five $1 bills in change. Each student took $1 back, leaving the rest, $2, as a tip for the server.

As she walked away from the cafe, Lin thought, "Wait—this doesn't make sense. Since I put in $10 and got $1 back, I wound up paying $9. So did Mai and Priya. Together, we paid $27. Then we left a $2 tip. That makes $29 total. And yet we originally gave the waiter $30. Where did the extra dollar go?"

Think about the situation and about Lin's question. Do you agree that the numbers didn't add up properly? Explain your reasoning.

Summary

Using Negative Numbers to Make Sense of Contexts

Sometimes we represent changes in a quantity with positive and negative numbers. If the quantity increases, the change is positive. If it decreases, the change is negative.

- Suppose 5 gallons of water is put in a washing machine. We can represent the change in the number of gallons as +5. If 3 gallons is emptied from the machine, we can represent the change as -3.

It is especially common to represent money we receive with positive numbers and money we spend with negative numbers.

- Suppose Clare gets $30.00 for her birthday and spends $18.00 buying lunch for herself and a friend. To her, the value of the gift can be represented as +30.00 and the value of the lunch as -18.00.

Whether a number is considered positive or negative depends on a person's perspective. If Clare's grandmother gives her $20 for her birthday, Clare might see this as +20, because to her, the amount of money she has increased. But her grandmother might see it as -20, because to her, the amount of money she has decreased.

In general, when using positive and negative numbers to represent changes, we have to be very clear about what it means when the change is positive and what it means when the change is negative.

NAME _____ DATE _____ PERIOD _____

Practice
Using Negative Numbers to Make Sense of Contexts

1. Write a positive or negative number to represent each change in the high temperature.

   a. Tuesday's high temperature was 4 degrees less than Monday's high temperature.

   b. Wednesday's high temperature was 3.5 degrees less than Tuesday's high temperature.

   c. Thursday's high temperature was 6.5 degrees more than Wednesday's high temperature.

   d. Friday's high temperature was 2 degrees less than Thursday's high temperature.

2. Decide which of the following quantities can be represented by a positive number and which can be represented by a negative number. Give an example of a quantity with the opposite sign in the same situation.

   a. Tyler's puppy gained 5 pounds.

   b. The aquarium leaked 2 gallons of water.

   c. Andre received a gift of $10.

   d. Kiran gave a gift of $10.

   e. A climber descended 550 feet.

3. Make up a situation where a quantity is changing.

    a. Explain what it means to have a negative change.

    b. Explain what it means to have a positive change.

    c. Give an example of each.

4. Respond to each of the following. **(Lesson 7-2)**

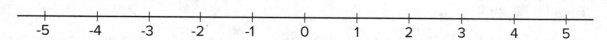

    a. On the number line, label the points that are 4 units away from 0.

    b. If you fold the number line so that a vertical crease goes through 0, the points you label would match up. Explain why this happens.

    c. On the number line, label the points that are $\frac{5}{2}$ units from 0. What is the distance between these points?

5. Evaluate each expression. **(Lesson 6-12)**

    a. $2^3 \cdot 3$ **b.** $\dfrac{4^2}{2}$

    c. 3^1 **d.** $6^2 \div 4$

    e. $2^3 - 2$ **f.** $10^2 + 5^2$

Lesson 7-6

Absolute Value of Numbers

NAME _____ DATE _____ PERIOD _____

Learning Goal Let's explore distances from zero more closely.

Warm Up
6.1 Number Talk: Closer to Zero

For each pair of expressions, decide mentally which one has a value that is closer to 0.

1. $\frac{9}{11}$ or $\frac{15}{11}$

2. $\frac{1}{5}$ or $\frac{1}{9}$

3. 1.25 or $\frac{5}{4}$

4. 0.01 or 0.001

Activity

6.2 Jumping Flea

1. A flea is jumping around on a number line.

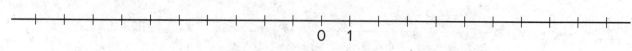

   a. If the flea starts at 1 and jumps 4 units to the right, where does it end up? How far away from 0 is this?

   b. If the flea starts at 1 and jumps 4 units to the left, where does it end up? How far away from 0 is this?

   c. If the flea starts at 0 and jumps 3 units away, where might it land?

   d. If the flea jumps 7 units and lands at 0, where could it have started?

   e. The **absolute value** of a number is the distance it is from 0. The flea is currently to the left of 0 and the absolute value of its location is 4. Where on the number line is it?

   f. If the flea is to the left of 0 and the absolute value of its location is 5, where on the number line is it?

   g. If the flea is to the right of 0 and the absolute value of its location is 2.5, where on the number line is it?

2. We use the notation $|\text{-}2|$ to say "the absolute value of -2," which means "the distance of -2 from 0 on the number line."

   a. What does $|\text{-}7|$ mean and what is its value?

   b. What does $|1.8|$ mean and what is its value?

NAME _____ DATE _____ PERIOD _____

 ## Activity
6.3 Absolute Elevation and Temperature

1. A part of the city of New Orleans is 6 feet below sea level. We can use "-6 feet" to describe its elevation, and "|-6| feet" to describe its vertical distance from sea level. In the context of elevation, what would each of the following numbers describe?

   a. 25 feet

   b. |25| feet

   c. -8 feet

   d. |-8| feet

2. The elevation of a city is different from sea level by 10 feet. Name the two elevations that the city could have.

3. We write "-5°C" to describe a temperature that is 5 degrees Celsius below freezing point and "5°C" for a temperature that is 5 degrees above freezing. In this context, what do each of the following numbers describe?

   a. 1°C

   b. -4°C

   c. |12|°C

   d. |-7|°C

4. Respond to each question.

   a. Which temperature is colder: -6°C or 3°C?

   b. Which temperature is closer to freezing temperature: -6°C or 3°C?

   c. Which temperature has a smaller absolute value? Explain how you know.

Are you ready for more?

At a certain time, the difference between the temperature in New York City and in Boston was 7 degrees Celsius. The difference between the temperature in Boston and in Chicago was also 7 degrees Celsius. Was the temperature in New York City the same as the temperature in Chicago? Explain your answer.

We compare numbers by comparing their positions on the number line: the one farther to the right is greater; the one farther to the left is less.

Sometimes we wish to compare which one is closer to or farther from 0. For example, we may want to know how far away the temperature is from the freezing point of 0°C, regardless of whether it is above or below freezing.

The **absolute value** of a number tells us its distance from 0.

The absolute value of -4 is 4, because -4 is 4 units to the left of 0. The absolute value of 4 is also 4, because 4 is 4 units to the right of 0. Opposites always have the same absolute value because they both have the same distance from 0.

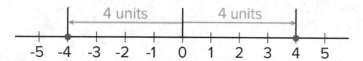

The distance from 0 to itself is 0, so the absolute value of 0 is 0. Zero is the *only* number whose distance to 0 is 0. For all other absolute values, there are always two numbers—one positive and one negative—that have that distance from 0.

To say "the absolute value of 4," we write: $|4|$.

To say that "the absolute value of -8 is 8," we write: $|-8| = 8$.

Glossary

absolute value

NAME _____ DATE _____ PERIOD _____

Practice
Absolute Value of Numbers

1. On the number line, plot and label all numbers with an absolute value of $\frac{3}{2}$.

```
  +----+----+----+----+
 -2   -1    0    1    2
```

2. The temperature at dawn is 6°C away from 0. Select **all** the temperatures that are possible.

   Ⓐ -12°C

   Ⓑ -6°C

   Ⓒ 0°C

   Ⓓ 6°C

   Ⓔ 12°C

3. Put these numbers in order, from least to greatest.

   $|-2.7|$ 0   1.3 $|-1|$ 2

4. Lin's family needs to travel 325 miles to reach her grandmother's house. **(Lesson 5-11)**

   a. At 26 miles, what percentage of the trip's distance have they completed?

   b. How far have they traveled when they have completed 72% of the trip's distance?

   c. At 377 miles, what percentage of the trip's distance have they completed?

5. Elena donates some money to charity whenever she earns money as a babysitter. The table shows how much money, d, she donates for different amounts of money, m, that she earns. (Lesson 6-16)

$ Donated , d	$ Earned, m
4.44	37
1.80	15
3.12	26
3.60	30
2.16	18

a. What percent of her income does Elena donate to charity? Explain or show your work.

b. Which quantity, m or d, would be the better choice for the dependent variable in an equation describing the relationship between m and d? Explain your reasoning.

c. Use your choice from the second question to write an equation that relates m and d.

6. How many times larger is the first number in the pair than the second? (Lesson 6-12)

a. 3^4 is _____ times larger than 3^3.

b. 5^3 is _____ times larger than 5^2.

c. 7^{10} is _____ times larger than 7^8.

d. 17^6 is _____ times larger than 17^4.

e. 5^{10} is _____ times larger than 5^4.

Lesson 7-7

Comparing Numbers and Distance from Zero

NAME _____ DATE _____ PERIOD _____

Learning Goal Let's use absolute value and negative numbers to think about elevation.

 Warm Up

7.1 Opposites

1. *a* is a rational number. Choose a value for *a* and plot it on the number line.

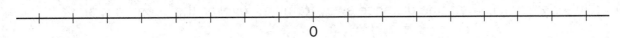

0

   a. Based on where you plotted *a*, plot -*a* on the same number line.

   b. What is the value of -*a* that you plotted?

2. Noah said, "If *a* is a rational number, -*a* will always be a negative number."
   Do you agree with Noah? Explain your reasoning.

Activity

7.2 Submarine

A submarine is at an elevation of -100 feet (100 feet below sea level). Let's compare the elevations of these four people to that of the submarine:

- Clare's elevation is greater than the elevation of the submarine. Clare is farther from sea level than the submarine.

- Andre's elevation is less than the elevation of the submarine. Andre is farther away from sea level than the submarine.

- Han's elevation is greater than the elevation of the submarine. Han is closer to sea level than is the submarine.

- Lin's elevation is the same distance away from sea level as the submarine's.

1. Complete the table as follows.

    a. Write a possible elevation for each person.

    b. Use <, >, or = to compare the elevation of that person to that of the submarine.

    c. Use absolute value to tell how far away the person is from sea level (elevation 0).

    As an example, the first row has been filled with a possible elevation for Clare.

	Possible Elevation	Compare to Submarine	Distance from Sea Level
Clare	150 feet	150 > -100	|150| or 150 feet
Andre			
Han			
Lin			

2. Priya says her elevation is less than the submarine's and she is closer to sea level. Is this possible? Explain your reasoning.

NAME _____ DATE _____ PERIOD _____

Activity

7.3 Info Gap: Points on the Number Line

Your teacher will give you either a *problem card* or a *data card*. Do not show or read your card to your partner.

If your teacher gives you the *problem card*:	If your teacher gives you the *data card*:
1. Silently read your card and think about what information you need to be able to answer the question. 2. Ask your partner for the specific information that you need. 3. Explain how you are using the information to solve the problem. Continue to ask questions until you have enough information to solve the problem. 4. Share the *problem card* and solve the problem independently. 5. Read the *data card* and discuss your reasoning.	1. Silently read your card. 2. Ask your partner *"What specific information do you need?"* and wait for them to *ask* for information. If your partner asks for information that is not on the card, do not do the calculations for them. Tell them you don't have that information. 3. Before sharing the information, ask *"Why do you need that information?"* Listen to your partner's reasoning and ask clarifying questions. 4. Read the *problem card* and solve the problem independently. 5. Share the *data card* and discuss your reasoning.

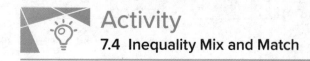
Here are some numbers and inequality symbols. Work with your partner to write true comparison statements.

-0.7	$\dfrac{7}{2}$	$<$
$-\dfrac{6}{3}$	4	$=$
-4	8	$>$
$-\dfrac{3}{5}$	$|3|$	
-2.5	$|-8|$	
0	$|0.7|$	
1	$\left|-\dfrac{5}{2}\right|$	
2.5		

One partner should select two numbers and one comparison symbol and use them to write a true statement using symbols. The other partner should write a sentence in words with the same meaning, using the following phrases:

- is equal to

- is the absolute value of

- is greater than

- is less than

For example, one partner could write $4 < 8$ and the other would write, "4 is less than 8." Switch roles until each partner has three true mathematical statements and three sentences written down.

NAME _____ DATE _____ PERIOD _____

For each question, choose a value for each variable to make the whole statement true. (When the word *and* is used in math, both parts have to be true for the whole statement to be true.) Can you do it if one variable is negative and one is positive? Can you do it if both values are negative?

1. $x < y$ and $|x| < y$.

2. $a < b$ and $|a| < |b|$.

3. $c < d$ and $|c| > d$.

4. $t < u$ and $|t| > |u|$.

Summary
Comparing Numbers and Distance from Zero

We can use elevation to help us compare two rational numbers or two absolute values.

- Suppose an anchor has an elevation of -10 meters and a house has an elevation of 12 meters. To describe the anchor having a lower elevation than the house, we can write -10 < 12 and say "-10 is less than 12."

- The anchor is closer to sea level than the house is to sea level (or elevation of 0). To describe this, we can write $|-10| < |12|$ and say "the distance between -10 and 0 is less than the distance between 12 and 0."

We can use similar descriptions to compare rational numbers and their absolute values outside of the context of elevation.

- To compare the distance of -47.5 and 5.2 from 0, we can say: $|-47.5|$ is 47.5 units away from 0, and $|5.2|$ is 5.2 units away from 0, so $|-47.5| > |5.2|$.

- $|-18| > 4$ means that the absolute value of -18 is greater than 4. This is true because 18 is greater than 4.

Practice
Comparing Numbers and Distance from Zero

1. In the context of elevation, what would $|-7|$ feet mean?

2. Match the statements written in English with the mathematical statements.

    a. The number -4 is a distance of 4 units away from 0 on the number line.

    b. The number -63 is more than 4 units away from 0 on the number line.

    c. The number 4 is greater than the number -4.

    d. The numbers 4 and -4 are the same distance away from 0 on the number line.

    e. The number -63 is less than the number 4.

    f. The number -63 is further away from 0 than the number 4 on the number line.

    Mathematical Statements

    1. $|-63| > 4$

    2. $-63 < 4$

    3. $|-63| > |4|$

    4. $|-4| = 4$

    5. $4 > -4$

    6. $|4| = |-4|$

3. Compare each pair of expressions using >, <, or =.

    a. -32 _____ 15

    b. $|-32|$ _____ $|15|$

    c. 5 _____ -5

    d. $|5|$ _____ $|-5|$

NAME _____ DATE _____ PERIOD _____

e. 2 _____ -17

f. 2 _____ |-17|

g. |-27| _____ |-45|

h. |-27| _____ -45

4. Mai received and spent money in the following ways last month.
For each example, write a signed number to represent the change
in money from her perspective. (Lesson 7- 5)

   a. Her grandmother gave her $25 in a birthday card.

   b. She earned $14 dollars babysitting.

   c. She spent $10 on a ticket to the concert.

   d. She donated $3 to a local charity.

   e. She got $2 interest on money that was in her savings account.

5. Here are the lowest temperatures recorded in the last 2 centuries for some US cities. (Lesson 7-1)

- Death Valley, CA was -45°F in January of 1937.
- Danbury, CT was -37°F in February of 1943.
- Monticello, FL was -2°F in February of 1899.
- East Saint Louis, IL was -36°F in January of 1999.
- Greenville, GA was -17°F in January of 1940.

a. Which of these states has the lowest record temperature?

b. Which state has a lower record temperature, FL or GA?

c. Which state has a lower record temperature, CT or IL?

d. How many more degrees colder is the record temperature for GA than for FL?

6. Find the quotients. (Lesson 5-13)

a. $0.024 \div 0.015$

b. $0.24 \div 0.015$

c. $0.024 \div 0.15$

d. $24 \div 15$

Lesson 7-8

Writing and Graphing Inequalities

NAME _____ DATE _____ PERIOD _____

Learning Goal Let's write inequalities.

Warm Up
8.1 Estimate Heights of People

1. Here is a picture of a man.

   a. Name a number, in feet, that is clearly too high for this man's height.

   b. Name a number, in feet, that is clearly too low for his height.

   c. Make an estimate of his height.

   Pause here for a class discussion.

2. Here is a picture of the same man standing next to a child.

   If the man's actual height is 5 feet 10 inches, what can you say about the height of the child in this picture?

   Be prepared to explain your reasoning.

Activity

8.2 Stories about 9

1. Your teacher will give you a set of paper slips with four stories and questions involving the number 9. Match each question to three representations of the solution: a description or a list, a number line, or an inequality statement.

2. Compare your matching decisions with another group's. If there are disagreements, discuss until both groups come to an agreement. Then, record your final matching decisions here.

   a. A fishing boat can hold fewer than 9 people. How many people (x) can it hold?

   - Description or list:

   - Number line:

   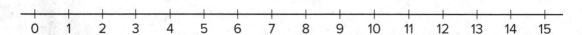

   - Inequality:

   b. Lin needs more than 9 ounces of butter to make cookies for her party. How many ounces of butter (x) would be enough?

   - Description or list:

   - Number line:

   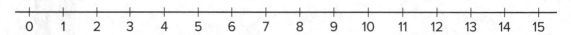

   - Inequality:

NAME _____ DATE _____ PERIOD _____

c. A magician will perform her magic tricks only if there are at least
9 people in the audience. For how many people (*x*) will she perform
her magic tricks?

- Description or list:

- Number line:

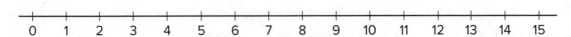

- Inequality:

d. A food scale can measure up to 9 kilograms of weight. What
weights (*x*) can the scale measure?

- Description or list:

- Number line:

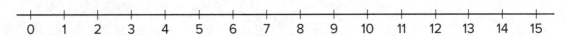

- Inequality:

Activity

8.3 How High and How Low Can It Be?

Here is a picture of a person and a basketball hoop. Based on the picture, what do you think are reasonable estimates for the maximum and minimum heights of the basketball hoop?

1. Complete the first blank in each sentence with an estimate, and the second blank with "taller" or "shorter."

   a. I estimate the *minimum* height of the basketball hoop to be _____ feet; this means the hoop cannot be_____ than this height.

   b. I estimate the *maximum* height of the basketball hoop to be _____ feet; this means the hoop cannot be _____ than this height.

2. Write two inequalities—one to show your estimate for the *minimum* height of the basketball hoop, and another for the *maximum* height. Use an inequality symbol and the variable h to represent the unknown height.

3. Plot each estimate for minimum or maximum value on a number line.

   • Minimum:

   • Maximum:

NAME _____ DATE _____ PERIOD _____

4. Suppose a classmate estimated the value of *h* to be 19 feet. Does this estimate agree with your inequality for the maximum height? Does it agree with your inequality for the minimum height? Explain or show how you know.

5. Ask a partner for an estimate of *h*. Record the estimate and check if it agrees with your inequalities for maximum and minimum heights.

Are you ready for more?

1. Find 3 different numbers that *a* could be if $|a| < 5$. Plot these points on the number line. Then plot as many other possibilities for *a* as you can.

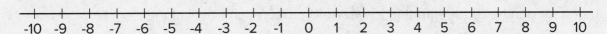

2. Find 3 different numbers that *b* could be if $|b| > 3$. Plot these points on the number line. Then plot as many other possibilities for *b* as you can.

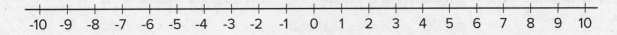

Summary
Writing and Graphing Inequalities

An inequality tells us that one value is *less than* or *greater than* another value.

Suppose we knew the temperature is *less than* 3°F, but we don't know exactly what it is. To represent what we know about the temperature t in °F we can write the inequality: $t < 3$.

The temperature can also be graphed on a number line. Any point to the left of 3 is a possible value for t. The open circle at 3 means that t cannot be *equal* to 3, because the temperature is *less than* 3.

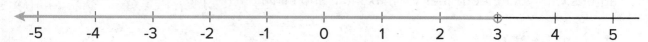

Here is another example.

Suppose a young traveler has to be at least 16 years old to fly on an airplane without an accompanying adult.

If a represents the age of the traveler, any number greater than 16 is a possible value for a, and 16 itself is also a possible value of a.

We can show this on a number line by drawing a closed circle at 16 to show that it meets the requirement (a 16-year-old person can travel alone).

From there, we draw a line that points to the right.

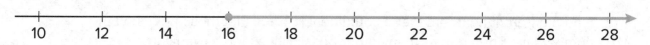

We can also write an inequality and equation to show possible values for a:

$$a > 16 \qquad\qquad a = 16$$

NAME _____ DATE _____ PERIOD _____

Practice
Writing and Graphing Inequalities

1. At the book sale, all books cost less than $5.

   a. What is the most expensive a book could be?

   b. Write an inequality to represent costs of books at the sale.

   c. Draw a number line to represent the inequality.

2. Kiran started his homework *before* 7:00 p.m. and finished his homework *after* 8:00 p.m. Let h represent the number of hours Kiran worked on his homework. Decide if each statement it is definitely true, definitely not true, or possibly true. Explain your reasoning.

   a. $h > 1$ b. $h > 2$

   c. $h < 1$ d. $h < 2$

3. Consider a rectangular prism with length 4 and width and height d. (Lesson 6-14)

   a. Find an expression for the volume of the prism in terms of d.

   b. Compute the volume of the prism when:

   • $d = 1$

   • $d = 2$

   • $d = \frac{1}{2}$

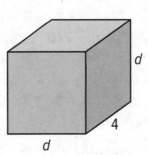

4. Match the statements written in English with the mathematical statements. All of these statements are true. **(Lesson 7-7)**

a. The number -15 is further away from 0 than the number -12 on the number line.

b. The number -12 is a distance of 12 units away from 0 on the number line.

c. The distance between -12 and 0 on the number line is greater than -15.

d. The numbers 12 and -12 are the same distance away from 0 on the number line.

e. The number -15 is less than the number -12.

f. The number 12 is greater than the number -12.

Mathematical Statements

1. $|-12| > -15$

2. $-15 < -12$

3. $|-15| > |-12|$

4. $|-12| = 12$

5. $12 > -12$

6. $|12| = |-12|$

5. Here are five sums. Use the distributive property to write each sum as a product with two factors. **(Lesson 6-11)**

a. $2a + 7a$

b. $5z - 10$

c. $c - 2cd$

d. $r + r + r + r$

e. $2x - \dfrac{1}{2}$

Lesson 7-9

Solutions of Inequalities

NAME _____ DATE _____ PERIOD _____

Learning Goal Let's think about the solutions to inequalities.

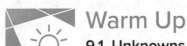 Warm Up
9.1 Unknowns on a Number Line

The number line shows several points, each labeled with a letter.

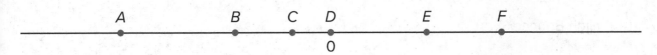

1. Fill in each blank with a letter so that the inequality statements are true.

   a. _____ > _____

   b. _____ < _____

2. Jada says that she found three different ways to complete the first question correctly. Do you think this is possible? Explain your reasoning.

3. List a possible value for each letter on the number line based on its location.

Activity

9.2 Amusement Park Rides

Priya finds these height requirements for some of the rides at an amusement park.

To ride the . . .	you must be . . .
High Bounce	between 55 and 72 inches tall
Climb-A-Thon	under 60 inches tall
Twirl-O-Coaster	58 inches minimum

1. Write an inequality for each of the three height requirements. Use *h* for the unknown height. Then, represent each height requirement on a number line.

   • High Bounce

   • Climb-A-Thon

   • Twirl-O-Coaster

NAME _____ DATE _____ PERIOD _____

2. Han's cousin is 55 inches tall. Han doesn't think she is tall enough to ride the High Bounce, but Kiran believes that she is tall enough. Do you agree with Han or Kiran? Be prepared to explain your reasoning.

Pause here for additional instructions from your teacher.

3. Priya can ride the Climb-A-Thon, but she cannot ride the High Bounce or the Twirl-O-Coaster. Which, if any, of the following could be Priya's height? Be prepared to explain your reasoning.

59 inches   53 inches   56 inches

4. Jada is 56 inches tall. Which rides can she go on?

5. Kiran is 60 inches tall. Which rides can he go on?

6. The inequalities $h < 75$ and $h > 64$ represent the height restrictions, in inches, of another ride. Write three values that are **solutions** to both of these inequalities.

1. Represent the height restrictions for all three rides on a single number line, using a different color for each ride.

2. Which part of the number line is shaded with all 3 colors?

3. Name one possible height a person could be in order to go on all three rides.

Activity

9.3 What Number Am I?

Your teacher will give your group two sets of cards—one set shows inequalities and the other shows numbers. Place the inequality cards face up where everyone can see them. Shuffle the number cards and stack them face down.

To play:

- One person in your group is the detective. The other people will give clues.

- Pick one number card from the stack and show it to everyone except the detective.

- The people giving clues each choose an inequality that will help the detective identify the unknown number.

- The detective studies the inequalities and makes three guesses.

  - If the detective does not guess the right number, each person chooses another inequality to help.

  - When the detective does guess the right number, a new person becomes the detective.

- Repeat the game until everyone has had a turn being the detective.

NAME _____ DATE _____ PERIOD _____

Summary
Solutions of Inequalities

Let's say a movie ticket costs less than $10. If c represents the cost of a movie ticket, we can use $c < \$10$ to express what we know about the cost of a ticket. Any value of c that makes the inequality true is called a **solution to the inequality**.

For example, 5 is a solution to the inequality $c < 10$ because $5 < 10$ (or "5 is less than 10") is a true statement, but 12 is not a solution because $12 < 10$ ("12 is less than 10") is *not* a true statement.

If a situation involves more than one boundary or limit, we will need more than one inequality to express it.

For example, if we knew that it rained for *more* than 10 minutes but *less* than 30 minutes, we can describe the number of minutes that it rained (r) with the following inequalities and number lines.

$r > 10$

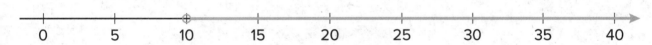

$r < 30$

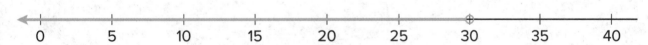

Any number of minutes greater than 10 is a solution to $r > 10$, and any number less than 30 is a solution to $r < 30$. But to meet the condition of "more than 10 but less than 30," the solutions are limited to the numbers between 10 and 30 minutes, *not* including 10 and 30. We can show the solutions visually by graphing the two inequalities on one number line.

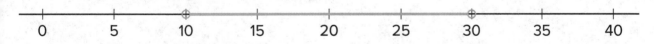

Glossary

solution to an inequality

1. Respond to each of the following.

   a. Select **all** numbers that are solutions to the inequality $k > 5$.

   (A.) 4

   (B.) 5

   (C.) 6

   (D.) 5.2

   (E.) 5.01

   (F.) 0.5

   b. Draw a number line to represent this inequality.

2. A sign on the road says: "Speed limit, 60 miles per hour."

   a. Let s be the speed of a car. Write an inequality that matches the information on the sign.

   b. Draw a number line to represent the solutions to the inequality.

   c. Could 60 be a value of s? Explain your reasoning.

NAME _____ DATE _____ PERIOD _____

3. One day in Boston, MA, the high temperature was 60 degrees Fahrenheit, and the low temperature was 52 degrees.

   a. Write one or more inequalities to describe the temperatures T that are between the high and low temperature on that day.

   b. Show the possible temperatures on a number line.

4. Select **all** the true statements. (Lesson 7-7)

   (A.) $-5 < |-5|$

   (B.) $|-6| < -5$

   (C.) $|-6| < 3$

   (D.) $4 < |-7|$

   (E.) $|-7| < |-8|$

5. Match each equation to its solution. (Lesson 6-15)

Equations	Solutions
a. $x^4 = 81$	2
b. $x^2 = 100$	3
c. $x^3 = 64$	4
d. $x^5 = 32$	10

6. Respond to each of the following. (Lesson 3-14)

    a. The price of a cell phone is $250. Elena's mom buys one of these cell phones for $150. What percentage of the usual price did she pay?

    b. Elena's dad buys another type of cell phone that also usually sells for $250. He pays 75% of the usual price. How much did he pay?

Lesson 7-10

Interpreting Inequalities

NAME _____ DATE _____ PERIOD _____

Learning Goal Let's examine what inequalities can tell us.

 ## Warm Up
10.1 True or False: Fractions and Decimals

Is each equation true or false? Be prepared to explain your reasoning.

1. $3(12 + 5) = (3 \cdot 12) \cdot (3 \cdot 5)$

2. $\dfrac{1}{3} \cdot \dfrac{3}{4} = \dfrac{3}{4} \cdot \dfrac{2}{6}$

3. $2 \cdot (1.5) \cdot 12 = 4 \cdot (0.75) \cdot 6$

 ## Activity
10.2 Basketball Game

Noah scored n points in a basketball game.

1. What does $15 < n$ mean in the context of the basketball game?

2. What does $n < 25$ mean in the context of the basketball game?

3. Draw two number lines to represent the solutions to the two inequalities.

4. Name a possible value for n that is a solution to both inequalities.

5. Name a possible value for n that is a solution to $15 < n$, but not a solution to $n < 25$.

6. Can -8 be a solution to n in this context? Explain your reasoning.

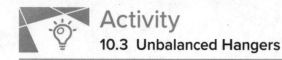
1. Here is a diagram of an unbalanced hanger.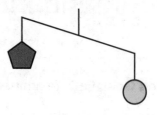

   a. Jada says that the weight of one circle is greater than the weight of one pentagon. Write an inequality to represent her statement. Let p be the weight of one pentagon and c be the weight of one circle.

   b. A circle weighs 12 ounces. Use this information to write another inequality to represent the relationship of the weights. Then, describe what this inequality means in this context.

2. Here is another diagram of an unbalanced hanger.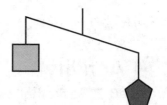

   a. Write an inequality to represent the relationship of the weights. Let p be the weight of one pentagon and s be the weight of one square.

   b. One pentagon weighs 8 ounces. Use this information to write another inequality to represent the relationship of the weights. Then, describe what this inequality means in this context.

   c. Graph the solutions to this inequality on a number line.

3. Based on your work so far, can you tell the relationship between the weight of a square and the weight of a circle? If so, write an inequality to represent that relationship. If not, explain your reasoning.

NAME _____ DATE _____ PERIOD _____

4. This is another diagram of an unbalanced hanger.
Andre writes the following inequality: $c + p < s$.
Do you agree with his inequality?
Explain your reasoning.

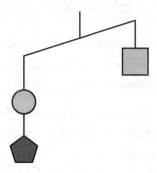

5. Jada looks at another diagram of an unbalanced hanger and writes:
$s + c > 2t$, where t represents the weight of one triangle. Draw a sketch
of the diagram.

Are you ready for more?

Here is a picture of a balanced hanger. It shows that the total
weight of the three triangles is the same as the total weight
of the four squares.

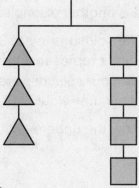

1. What does this tell you about the weight of one square
when compared to one triangle? Explain how you know.

2. Write an equation or an inequality to describe the relationship
between the weight of a square and that of a triangle. Let s be
the weight of a square and t be the weight of a triangle.

When we find the solutions to an inequality, we should think about its context carefully. A number may be a solution to an inequality outside of a context, but may not make sense when considered in context.

- Suppose a basketball player scored more than 11 points in a game, and we represent the number of points she scored, s, with the inequality $s > 11$.

  By looking only at $s > 11$, we can say that numbers such as 12, $14\frac{1}{2}$, and 130.25 are all solutions to the inequality because they each make the inequality true.

$$12 > 11$$

$$14\frac{1}{2} > 11$$

$$130.25 > 11$$

  In a basketball game, however, it is only possible to score a whole number of points, so fractional and decimal scores are not possible. It is also highly unlikely that one person would score more than 130 points in a single game.

  In other words, the context of an inequality may limit its solutions.

Here is another example:

- The solutions to $r < 30$ can include numbers such as $27\frac{3}{4}$, 18.5, 0, and -7. But if r represents the number of minutes of rain yesterday (and it did rain), then our solutions are limited to positive numbers. Zero or negative number of minutes would not make sense in this context.

To show the upper and lower boundaries, we can write two inequalities:

$$0 < r$$

$$r < 30$$

Inequalities can also represent comparison of two unknown numbers.

- Let's say we knew that a puppy weighs more than a kitten, but we did not know the weight of either animal. We can represent the weight of the puppy, in pounds, with p and the weight of the kitten, in pounds, with k, and write this inequality:

$$p > k$$

NAME _____ DATE _____ PERIOD _____

Practice
Interpreting Inequalities

1. There is a closed carton of eggs in Mai's refrigerator. The carton contains *e* eggs and it can hold 12 eggs.

   a. What does the inequality $e < 12$ mean in this context?

   b. What does the inequality $e > 0$ mean in this context?

   c. What are some possible values of *e* that will make both $e < 12$ and $e > 0$ true?

2. Here is a diagram of an unbalanced hanger.

   a. Write an inequality to represent the relationship of the weights. Use *s* to represent the weight of the square in grams and *c* to represent the weight of the circle in grams.

   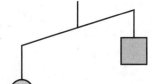

   b. One red circle weighs 12 grams. Write an inequality to represent the weight of one blue square.

   c. Could 0 be a value of *s*? Explain your reasoning.

3. Respond to each of the following. **(Lesson 7-8)**

   a. Jada is taller than Diego. Diego is 54 inches tall (4 feet, 6 inches). Write an inequality that compares Jada's height in inches, *j*, to Diego's height.

   b. Jada is shorter than Elena. Elena is 5 feet tall. Write an inequality that compares Jada's height in inches, *j*, to Elena's height.

4. Tyler has more than $10. Elena has more money than Tyler. Mai has more money than Elena. Let t be the amount of money that Tyler has, let e be the amount of money that Elena has, and let m be the amount of money that Mai has. Select **all** statements that are true:

(A.) $t < j$

(B.) $m > 10$

(C.) $e > 10$

(D.) $t > 10$

(E.) $e > m$

(F.) $t < e$

5. Which is greater, $\frac{-9}{20}$ or -0.5? Explain how you know. If you get stuck, consider plotting the numbers on a number line. **(Lesson 7-3)**

6. Select **all** the expressions that are equivalent to $\left(\frac{1}{2}\right)^3$. **(Lesson 6-13)**

(A.) $\frac{1}{2} \cdot \frac{1}{2} \cdot \frac{1}{2}$

(B.) $\frac{1}{2^3}$

(C.) $\left(\frac{1}{3}\right)^2$

(D.) $\frac{1}{6}$

(E.) $\frac{1}{8}$

Lesson 7-11

Points on the Coordinate Plane

NAME _____ DATE _____ PERIOD _____

Learning Goal Let's explore and extend the coordinate plane.

Warm Up
11.1 Guess My Line

1. Choose a horizontal or a vertical line on the grid. Draw 4 points on the line and label each point with its coordinates.

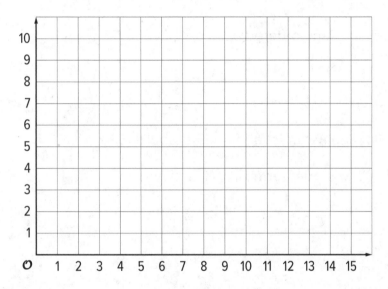

2. Tell your partner whether your line is horizontal or vertical, and have your partner guess the locations of your points by naming coordinates.

   If a guess is correct, put an X through the point. If your partner guessed a point that is on your line but not the point that you plotted, say, "That point is on my line, but is not one of my points."

   Take turns guessing each other's points, 3 guesses per turn.

1. Label each point on the coordinate plane with an ordered pair.

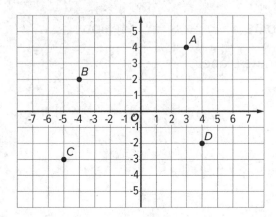

2. What do you notice about the locations and ordered pairs of *B*, *C*, and *D*? How are they different from those for point *A*?

3. Plot a point at (-2, 5). Label it *E*. Plot another point at (3, -4.5). Label it *F*.

4. The coordinate plane is divided into four **quadrants**, I, II, III, and IV, as shown here.

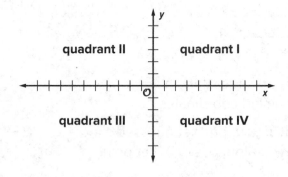

$G = (5, 2)$ $H = (-1, -5)$ $I = (7, -4)$

a. In which quadrant is:
   point *G* located?

   point *H* located?

   point *I* located?

b. A point has a positive *y*-coordinate. In which quadrant could it be?

NAME _____ DATE _____ PERIOD _____

Activity

11.3 Coordinated Archery

Here is an image of an archery target on a coordinate plane. The scores for landing an arrow in the colored regions are shown.

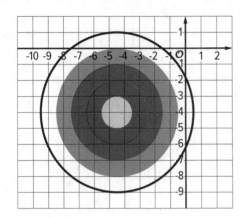

- Yellow: 10 points
- Red: 8 points
- Blue: 6 points
- Green: 4 points
- White: 2 points

Name the coordinates for a possible landing point to score.

1. 6 points

2. 10 points

3. 2 points

4. No points

5. 4 points

6. 8 points

Are you ready for more?

Pretend you are stuck in a coordinate plane. You can only take vertical and horizontal steps that are one unit long.

1. How many ways are there to get from the point (-3, 2) to (-1, -1) if you will only step down and to the right?

2. How many ways are there to get from the point (-1, -2) to (4, 0) if you can only step up and to the right?

3. Make up some more problems like this and see what patterns you notice.

Just as the number line can be extended to the left to include negative numbers, the x- and y-axis of a coordinate plane can also be extended to include negative values.

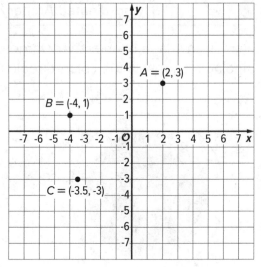

The ordered pair (x, y) can have negative x- and y-values. For $B = (-4, 1)$, the x-value of -4 tells us that the point is 4 units to the left of the y-axis. The y-value of 1 tells us that the point is one unit above the x-axis.

The same reasoning applies to the points A and C. The x- and y-coordinates for point A are positive, so A is to the right of the y-axis and above the x-axis. The x- and y-coordinates for point C are negative, so C is to the left of the y-axis and below the x-axis.

Glossary

quadrant

NAME _____ DATE _____ PERIOD _____

Practice
Points on the Coordinate Plane

1. Respond to each of the following.

   a. Graph these points on the coordinate plane: (-2,3), (2,3), (-2,-3), (2,-3).

   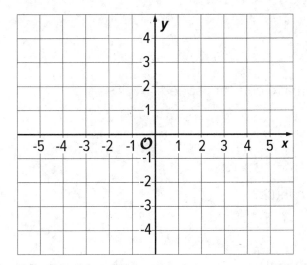

   b. Connect all of the points. Describe the figure.

2. Write the coordinates of each point.

   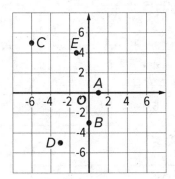

3. These three points form a horizontal line: (-3.5,4), (0,4), and (6.2,4).
   Name two additional points that fall on this line.

4. One night, it is 24°C warmer in Tucson than it was in Minneapolis. If the temperatures in Tucson and Minneapolis are opposites, what is the temperature in Tucson? (Lesson 7-2)

(A.) -24°C

(B.) -12°C

(C.) 12°C

(D.) 24°C

5. Lin ran 29 meters in 10 seconds. She ran at a constant speed. (Lesson 2-9)

   a. How far did Lin run every second?

   b. At this rate, how far can she run in 1 minute?

6. Noah is helping his band sell boxes of chocolate to fund a field trip. Each box contains 20 bars and each bar sells for $1.50. (Lesson 6-16)

   a. Complete the table for values of *m*.

   b. Write an equation for the amount of money, *m*, that will be collected if *b* boxes of chocolate bars are sold. Which is the independent variable and which is the dependent variable in your equation?

   c. Write an equation for the number of boxes, *b*, that were sold if *m* dollars were collected. Which is the independent variable and which is the dependent variable in your equation?

Boxes Sold (*b*)	Money Collected (*m*)
1	
2	
3	
4	
5	
6	
7	
8	

Lesson 7-12

Constructing the Coordinate Plane

NAME _____ DATE _____ PERIOD _____

Learning Goal Let's investigate different ways of creating a coordinate plane.

Warm Up
12.1 English Winter

The following data were collected over one December afternoon in England.

Time after Noon (hours)	Temperature (°C)
0	5
1	3
2	4
3	2
4	1
5	-2
6	-3
7	-4
8	-4

A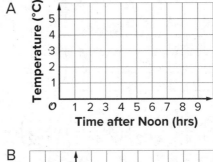

1. Which set of axes would you choose to represent these data? Explain your reasoning.

B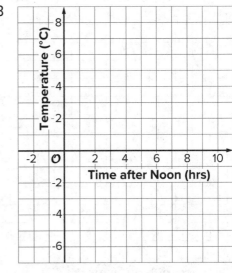

2. Explain why the other two sets of axes did not seem as appropriate as the one you chose.

C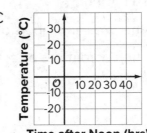

1. Here are three sets of coordinates. For each set, draw and label an appropriate pair of axes and plot the points.

   a. (1, 2), (3, -4), (-5, -2), (0, 2.5)

   b. (50, 50), (0, 0), (-10, -30), (-35, 40)

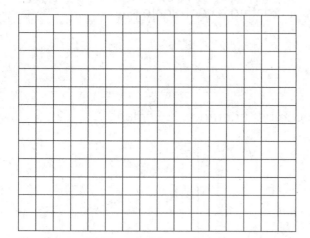

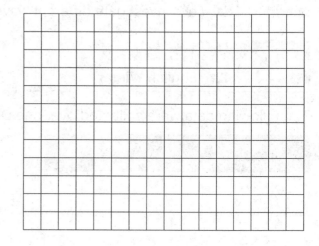

   c. $\left(\frac{1}{4}, \frac{3}{4}\right), \left(\frac{-5}{4}, \frac{1}{2}\right), \left(-1\frac{1}{4}, \frac{-3}{4}\right), \left(\frac{1}{4}, \frac{-1}{2}\right)$

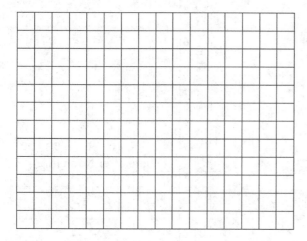

2. Discuss with a partner:

   a. How are the axes and labels of your three drawings different?

   b. How did the coordinates affect the way you drew the axes and label the numbers?

NAME _____ DATE _____ PERIOD _____

Activity
12.3 Positively A-maze-ing

Here is a maze on a coordinate plane. The black point in the center is (0, 0). The side of each grid square is 2 units long.

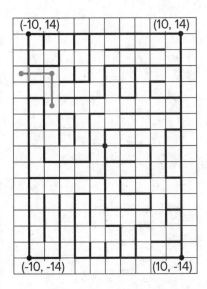

1. Enter the above maze at the location marked with a green segment. Draw line segments to show your way through and out of the maze. Label each turning point with a letter. Then, list all the letters and write their coordinates.

2. Choose any 2 turning points that share the same line segment. What is the same about their coordinates? Explain why they share that feature.

To get from the point (2, 1) to (-4, 3) you can go two units up and six units to the left, for a total distance of eight units. This is called the "taxicab distance," because a taxi driver would have to drive eight blocks to get between those two points on a map.

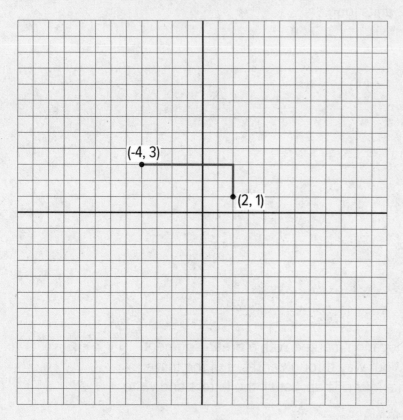

Find as many points as you can that have a taxicab distance of eight units away from (2,1). What shape do these points make?

NAME _____ DATE _____ PERIOD _____

Summary
Constructing the Coordinate Plane

The coordinate plane can be used to show information involving pairs of numbers. When using the coordinate plane, we should pay close attention to what each axis represents and what scale each uses.

Suppose we want to plot the following data about the temperatures in Minneapolis one evening.

We can decide that the x-axis represents number of hours in relation to midnight and the y-axis represents temperatures in degrees Celsius.

Time (hours from midnight)	Temperature (degrees C)
-4	3
-1	-2
0	-4
3	-8

- In this case, x-values less than 0 represent hours before midnight, and x-values greater than 0 represent hours after midnight.

- On the y-axis, the values represents temperatures above and below the freezing point of 0 degrees Celsius.

The data involve whole numbers, so it is appropriate that each square on the grid represents a whole number.

- On the left of the origin, the x-axis needs to go as far as -4 or less (farther to the left). On the right, it needs to go to 3 or greater.

- Below the origin, the y-axis has to go as far as -8 or lower. Above the origin, it needs to go to 3 or higher.

Here is a graph of the data with the axes labeled appropriately.

On this coordinate plane, a point at (0, 0) would mean a temperature of 0 degrees Celsius at midnight. The point at (-4, 3) means a temperature of 3 degrees Celsius at 4 hours before midnight (or 8 p.m.).

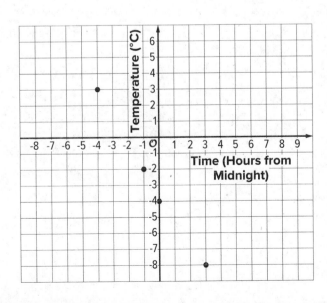

Practice

Constructing the Coordinate Plane

1. In the space provided, draw and label an appropriate pair of axes and plot the points.

   a. $\left(\dfrac{1}{5}, \dfrac{4}{5}\right)$

   b. $\left(-\dfrac{3}{5}, \dfrac{2}{5}\right)$

   c. $\left(-1\dfrac{1}{5}, -\dfrac{4}{5}\right)$

   d. $\left(\dfrac{1}{5}, -\dfrac{3}{5}\right)$

2. Diego was asked to plot these points: (-50, 0), (150, 100), (200, -100), (350, 50), (-250, 0). What interval could he use for each axis? Explain your reasoning.

NAME _____ DATE _____ PERIOD _____

3. Respond to each of the following.

   a. Name 4 points that would form a square with the origin at its center.

   b. Graph these points to check if they form a square.

4. Which of the following changes would you represent using a negative number? Explain what a positive number would represent in that situation. **(Lesson 7-5)**

   a. A loss of 4 points

   b. A gain of 50 yards

   c. A loss of $10

   d. An elevation above sea level

5. Jada is buying notebooks for school. The cost of each notebook is $1.75. (Lesson 6-16)

   a. Write an equation that shows the cost of Jada's notebooks, c, in terms of the number of notebooks, n, that she buys.

   b. Which of the following could be points on the graph of your equation? Select **all** that apply.

   A. (1.75, 1)

   B. (2, 3.50)

   C. (5, 8.75)

   D. (17.50, 10)

   E. (9, 15.35)

6. A corn field has an area of 28.6 acres. It requires about 15,000,000 gallons of water. About how many gallons of water per acre is that? (Lesson 5-13)

   A. 5,000

   B. 50,000

   C. 500,000

   D. 5,000,000

Lesson 7-13

Interpreting Points on the Coordinate Plane

NAME _____ DATE _____ PERIOD _____

Learning goal Let's examine what points on the coordinate plane can tell us.

Warm Up
13.1 Unlabeled Points

Label each point on the coordinate plane with the appropriate letter and ordered pair.

$A = (7, -5.5)$ $B = (-8, 4)$

$C = (3, 2)$ $D = (-3.5, 0.2)$

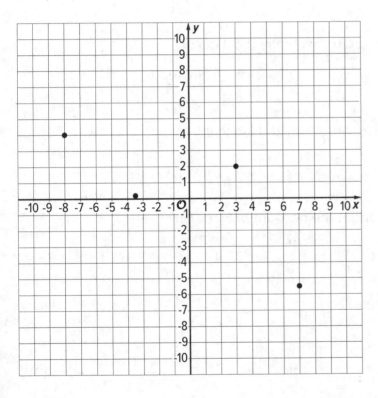

Activity

13.2 Account Balance

The graph shows the balance in a bank account over a period of 14 days. The axis labeled *b* represents account balance in dollars. The axis labeled *d* represents the day.

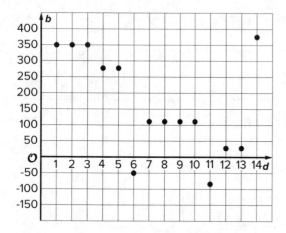

1. Estimate the greatest account balance. On which day did it occur?

2. Estimate the least account balance. On which day did it occur?

3. What does the point (6,-50) tell you about the account balance?

4. How can we interpret |-50| in the context?

NAME _____ DATE _____ PERIOD _____

Activity
13.3 High and Low Temperatures

The coordinate plane shows the high and low temperatures in Nome, Alaska over a period of 8 days. The axis labeled *T* represents temperatures in degrees Fahrenheit. The axis labeled *d* represents the day.

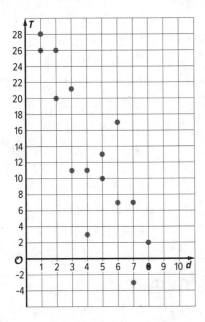

1. Respond to each of the following.

   a. What was the warmest high temperature?

   b. Write an inequality to describe the high temperatures, *H*, over the 8-day period.

   c. What was the coldest low temperature?

   d. Write an inequality to describe the low temperatures, *L*, over the 8-day period.

   e. On which day(s) did the *largest* difference between the high and low temperatures occur? Write down this difference.

   f. On which day(s) did the *smallest* difference between the high and low temperatures occur? Write down this difference.

Before doing this problem, do the problem about taxicab distance in an earlier lesson.

The point (0, 3) is 4 taxicab units away from (-4, 3) and 4 taxicab units away from (2, 1).

1. Find as many other points as you can that are 4 taxicab units away from *both* (-4, 3) and (2, 1).

2. Are there any points that are 3 taxicab units away from both points?

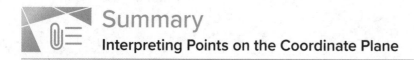

Summary

Interpreting Points on the Coordinate Plane

Points on the coordinate plane can give us information about a context or a situation. One of those contexts is about money.

To open a bank account, we have to put money into the account. The account balance is the amount of money in the account at any given time. If we put in $350 when opening the account, then the account balance will be 350.

Sometimes we may have no money in the account and need to borrow money from the bank. In that situation, the account balance would have a negative value. If we borrow $200, then the account balance is -200.

A coordinate grid can be used to display both the balance and the day or time for any balance. This allows to see how the balance changes over time or to compare the balances of different days.

Similarly, if we plot on the coordinate plane data such as temperature over time, we can see how temperature changes over time or compare temperatures of different times.

NAME _____ DATE _____ PERIOD _____

Practice
Interpreting Points on a Coordinate Plane

1. The elevation of a submarine is shown in the table. Draw and label coordinate axes with an appropriate scale and plot the points.

Time after Noon (hours)	Elevation (meters)
0	-567
1	-892
2	-1,606
3	-1,289
4	-990
5	-702
6	-365

2. The inequalities $h > 42$ and $h < 60$ represent the height requirements for an amusement park ride, where h represents a person's height in inches. Write a sentence or draw a sign that describes these rules as clearly as possible. **(Lesson 7-8)**

3. The *x*-axis represents the number of hours before or after noon, and the *y*-axis represents the temperature in degrees Celsius.

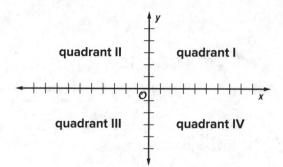

a. At 9 a.m., it was below freezing. In what quadrant would this point be plotted?

b. At 11 a.m., it was 10° C. In what quadrant would this point be plotted?

c. Choose another time and temperature. Then tell the quadrant where the point should be plotted.

d. What does the point (0, 0) represent in this context?

4. Solve each equation. (Lesson 6-4)

a. $3a = 12$

b. $b + 3.3 = 8.9$

c. $1 = \frac{1}{4}c$

d. $5\frac{1}{2} = d + \frac{1}{4}$

e. $2e = 6.4$

Lesson 7-14

Distances on a Coordinate Plane

NAME _____ DATE _____ PERIOD _____

Learning Goal Let's explore distance on the coordinate plane.

Warm Up
14.1 Coordinate Patterns

Plot points in your assigned quadrant and label them with their coordinates.

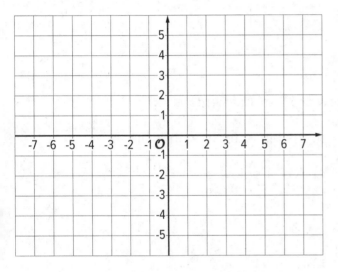

1. Write the coordinates of each point.

   A =

   B =

   C =

   D =

   E =

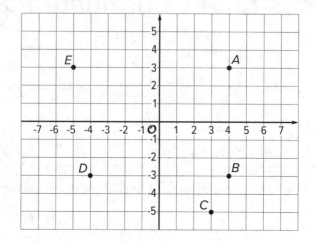

2. Answer these questions for each pair of points.

   • How are the coordinates the same? How are they different?

   • How far away are they from the *y*-axis? To the left or to the right of it?

   • How far away are they from the *x*-axis? Above or below it?

   a. *A* and *B*

   b. *B* and *D*

   c. *A* and *D*

Pause here for a class discussion.

NAME _____ DATE _____ PERIOD _____

3. Point *F* has the same coordinates as point *C*, except its *y*-coordinate has the opposite sign.

   a. Plot point *F* on the coordinate plane and label it with its coordinates.

   b. How far away are *F* and *C* from the *x*-axis?

   c. What is the distance between *F* and *C*?

4. Point *G* has the same coordinates as point *E*, except its *x*-coordinate has the opposite sign.

   a. Plot point *G* on the coordinate plane and label it with its coordinates.

   b. How far away are *G* and *E* from the *y*-axis?

   c. What is the distance between *G* and *E*?

5. Point *H* has the same coordinates as point *B*, except both of its coordinates have the opposite sign. In which quadrant is point *H*?

Activity

14.3 Finding Distances on a Coordinate Plane

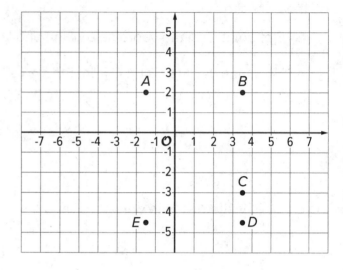

1. Label each point with its coordinates.

2. Find the distance between each of the following pairs of points.

   a. Point B and C

   b. Point D and B

   c. Point D and E

3. Which of the points are 5 units from (-1.5, -3)?

4. Which of the points are 2 units from (0.5, -4.5)?

5. Plot a point that is both 2.5 units from A and 9 units from E. Label that point M and write down its coordinates.

Are you ready for more?

Priya says, "There are exactly four points that are 3 units away from (-5, 0)." Lin says, "I think there are a whole bunch of points that are 3 units away from (-5, 0)." Do you agree with either of them? Explain your reasoning.

NAME _____ DATE _____ PERIOD _____

Summary

Distances on a Coordinate Plane

The points $A = (5, 2)$, $B = (-5, 2)$, $C = (-5, -2)$, and $D = (5, -2)$ are shown in the plane. Notice that they all have almost the same coordinates, except the signs are different. They are all the same distance from each axis but are in different quadrants.

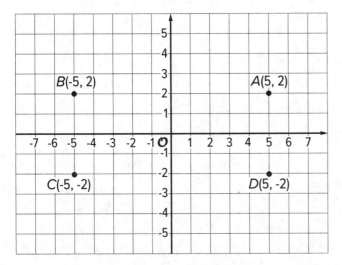

Notice that the vertical distance between points A and D is 4 units, because point A is 2 units above the horizontal axis and point D is 2 units below the horizontal axis. The horizontal distance between points A and B is 10 units, because point B is 5 units to the left of the vertical axis and point A is 5 units to the right of the vertical axis.

We can always tell which quadrant a point is located in by the signs of its coordinates.

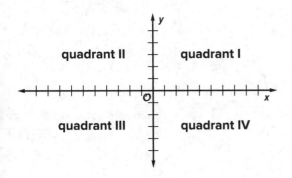

x	y	Quadrant
positive	positive	I
negative	positive	II
negative	negative	III
positive	negative	IV

In general:

- If two points have *x*-coordinates that are opposites (like 5 and -5), they are the same distance away from the vertical axis, but one is to the left and the other to the right.

- If two points have *y*-coordinates that are opposites (like 2 and -2), they are the same distance away from the horizontal axis, but one is above and the other below.

When two points have the same value for the first or second coordinate, we can find the distance between them by subtracting the coordinates that are different.

For example, consider (1, 3) and (5, 3):

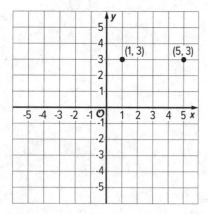

They have the same *y*-coordinate. If we subtract the *x*-coordinates, we get $5 - 1 = 4$. These points are 4 units apart.

NAME _____ DATE _____ PERIOD _____

Practice
Distances on a Coordinate Plane

1. Here are 4 points on a coordinate plane.

   a. Label each point with its coordinates.

   b. Plot a point that is 3 units from point *K*. Label it *P*.

   c. Plot a point that is 2 units from point *M*. Label it *W*.

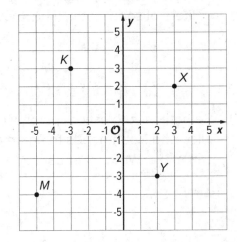

2. Each set of points are connected to form a line segment. What is the length of each?

   a. *A* = (3, 5) and *B* = (3, 6)

   b. *C* = (-2, -3) and *D* = (-2, -6)

   c. *E* = (-3, 1) and *F* = (-3, -1)

3. On the coordinate plane, plot four points that are each 3 units away from point *P* = (-2, -1). Write the coordinates of each point.

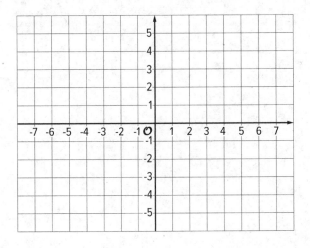

4. Noah's recipe for sparkling orange juice uses 4 liters of orange juice and 5 liters of soda water. (Lesson 6-16)

   a. Noah prepares large batches of sparkling orange juice for school parties. He usually knows the total number of liters, t, that he needs to prepare. Write an equation that shows how Noah can find s, the number of liters of soda water, if he knows t.

   b. Sometimes the school purchases a certain number, j, of liters of orange juice and Noah needs to figure out how much sparkling orange juice he can make. Write an equation that Noah can use to find t if he knows j.

5. For a suitcase to be checked on a flight (instead of carried by hand), it can weigh at most 50 pounds. Andre's suitcase weighs 23 kilograms. Can Andre check his suitcase? Explain or show your reasoning. (Note: 10 kilograms $\approx$ 22 pounds) (Lesson 3-4)

Lesson 7-15

Shapes on the Coordinate Plane

NAME _____ DATE _____ PERIOD _____

Learning Goal Let's use the coordinate plane to solve problems and puzzles.

 Warm Up

15.1 Figuring Out The Coordinate Plane

1. Draw a figure in the coordinate plane with at least three of following properties:

   • 6 vertices

   • Exactly 1 pair of parallel sides

   • At least 1 right angle

   • 2 sides with the same length

2. Is your figure a polygon? Explain how you know.

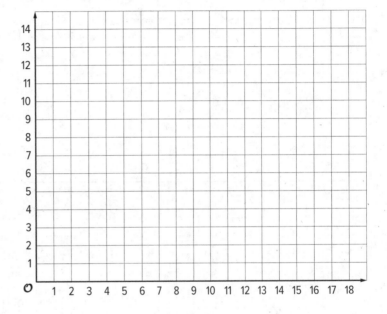

Activity

15.2 Plotting Polygons

Here are the coordinates for four polygons. Plot them on the coordinate plane, connect the points in the order that they are listed, and label each polygon with its letter name.

1. Polygon A: (-7, 4), (-8, 5), (-8, 6), (-7, 7), (-5, 7), (-5, 5), (-7, 4)

2. Polygon B: (4, 3), (3, 3), (2, 2), (2, 1), (3, 0), (4, 0), (5, 1), (5, 2), (4, 3)

3. Polygon C: (-8, -5), (-8, -8), (-5, -8), (-5, -5), (-8, -5)

4. Polygon D: (-5,1), (-3, -3), (-1, -2), (0, 3), (-3, 3), (-5, 1)

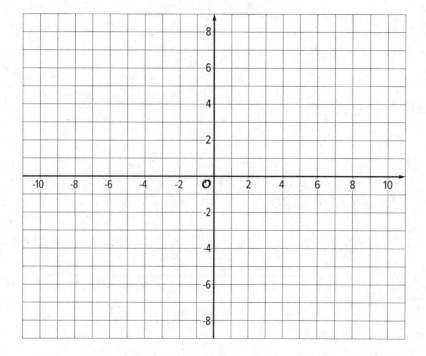

Are you ready for more?

Find the area of Polygon D in this activity.

NAME _____ DATE _____ PERIOD _____

 Activity

15.3 Four Quadrants of A-Maze-ing

The following diagram shows Andre's route through a maze. He started from the lower right entrance.

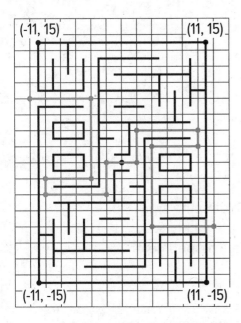

1. Respond to each of the following.

   a. What are the coordinates of the first two and the last two points of his route?

   b. How far did he walk from his starting point to his ending point? Show how you know.

2. Jada went into the maze and stopped at (-7,2).

   a. Plot that point and other points that would lead her out of the maze (through the exit on the upper left side).

   b. How far from (-7,2) must she walk to exit the maze? Show how you know.

We can use coordinates to find lengths of segments in the coordinate plane.

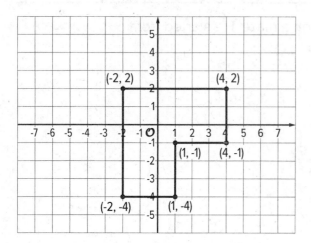

For example, we can find the perimeter of this polygon by finding the sum of its side lengths. Starting from (-2, 2) and moving clockwise, we can see that the lengths of the segments are 6, 3, 3, 3, 3, and 6 units. The perimeter is therefore 24 units.

In general:

- If two points have the same *x*-coordinate, they will be on the same vertical line, and we can find the distance between them.

- If two points have the same *y*-coordinate, they will be on the same horizontal line, and we can find the distance between them.

NAME _____ DATE _____ PERIOD _____

Practice

Shapes on the Coordinate Plane

1. The coordinates of a rectangle are (3, 0), (3, -5), (-4, 0) and (-4, -5)

   a. What is the length and width of this rectangle?

   b. What is the perimeter of the rectangle?

   c. What is the area of the rectangle?

2. Draw a square with one vertex on the point (-3, 5) and a perimeter of 20 units. Write the coordinates of each other vertex.

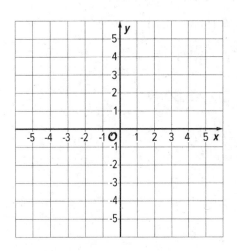

3. a. Plot and connect the following points to form a polygon: (-3, 2), (2, 2), (2, -4), (-1, -4), (-1, -2), (-3, -2), (-3, 2)

   b. Find the perimeter of the polygon.

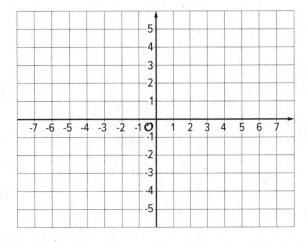

4. For each situation, select **all** the equations that represent it. Choose one equation and solve it. (Lesson 6-4)

   a. Jada's cat weighs 3.45 kg. Andre's cat weighs 1.2 kg more than Jada's cat. How much does Andre's cat weigh?

   $x = 3.45 + 1.2$

   $x = 3.45 - 1.2$

   $x + 1.2 = 3.45$

   $x - 1.2 = 3.45$

   b. Apples cost $1.60 per pound at the farmer's market. They cost 1.5 times as much at the grocery store. How much do the apples cost per pound at the grocery store?

   $y = (1.5) \cdot (1.60)$

   $y = 1.60 \div 1.5$

   $(1.5)y = 1.60$

   $\dfrac{y}{1.5} = 1.60$

Lesson 7-16

Common Factors

NAME _____ DATE _____ PERIOD _____

Learning Goal Let's use factors to solve problems.

Warm Up
16.1 Figures Made of Squares

How are the pairs of figures alike? How are they different?

Diego is preparing brownies and cookies for a bake sale. He would like to make equal-size bags for selling all of the 48 brownies and 64 cookies that he has. Organize your answer to each question so that it can be followed by others.

1. How can Diego package all the 48 brownies so that each bag has the same number of them? How many bags can he make, and how many brownies will be in each bag? Find all the possible ways to package the brownies.

2. How can Diego package all the 64 cookies so that each bag has the same number of them? How many bags can he make, and how many cookies will be in each bag? Find all the possible ways to package the cookies.

3. How can Diego package all the 48 brownies and 64 cookies so that each bag has the same combination of items? How many bags can he make, and how many of each will be in each bag? Find all the possible ways to package both items.

4. What is the largest number of combination bags that Diego can make with no left over? Explain to your partner how you know that it is the largest possible number of bags.

NAME _____ DATE _____ PERIOD _____

Activity
16.3 Greatest Common Factor

1. The **greatest common factor** of 30 and 18 is 6. What do you think the term "greatest common factor" means?

2. Find all of the **factors** of 21 and 6. Then, identify the greatest common factor of 21 and 6.

3. Find all of the factors of 28 and 12. Then, identify the greatest common factor of 28 and 12.

4. A rectangular bulletin board is 12 inches tall and 27 inches wide. Elena plans to cover it with squares of colored paper that are all the same size. The paper squares come in different sizes; all of them have whole-number inches for their side lengths.

   a. What is the side length of the largest square that Elena could use to cover the bulletin board completely without gaps and overlaps? Explain or show your reasoning.

   b. How is the solution to this problem related to greatest common factor?

Are you ready for more?

A school has 1,000 lockers, all lined up in a hallway. Each locker is closed. Then...

• One student goes down the hall and opens each locker.

• A second student goes down the hall and closes every second locker: lockers 2, 4, 6, and so on.

• A third student goes down the hall and changes every third locker. If a locker is open, he closes it. If a locker is closed, he opens it.

• A fourth student goes down the hall and changes every fourth locker.

This process continues up to the thousandth student! At the end of the process, which lockers will be open? (Hint: you may want to try this problem with a smaller number of lockers first.)

NAME _____ DATE _____ PERIOD _____

Summary
Common Factors

A factor of a whole number *n* is a whole number that divides *n* evenly without a remainder. For example, 1, 2, 3, 4, 6, and 12 are all factors of 12 because each of them divides 12 evenly and without a remainder.

A **common factor** of two whole numbers is a factor that they have in common. For example, 1, 3, 5, and 15 are factors of 45; they are also factors of 60. We call 1, 3, 5, and 15 common factors of 45 and 60.

The **greatest common factor** (sometimes written as GCF) of two whole numbers is the greatest of all of the common factors. For example, 15 is the greatest common factor for 45 and 60.

One way to find the greatest common factor of two whole numbers is to list all of the factors for each, and then look for the greatest factor they have in common. Let's try to find the greatest common factor of 18 and 24. First, we list all the factors of each number.

- Factors of 18: **1**, **2**, **3**, **6**, 9, 18
- Factors of 24: **1**, **2**, **3**, 4, **6**, 8, 12, 24

The common factors are 1, 2, 3, and 6. Of these, 6 is the greatest one, so 6 is the greatest common factor of 18 and 24.

> **Glossary**
>
> **common factor**
> **greatest common factor**

Practice
Common Factors

1. A teacher is making gift bags. Each bag is to be filled with pencils and stickers. The teacher has 24 pencils and 36 stickers to use. Each bag will have the same number of each item, with no items left over. For example, she could make 2 bags with 12 pencils and 18 stickers each. What are the other possibilities? Explain or show your reasoning.

2. Respond to each of the following.

    a. List all the factors of 42.

    b. What is the greatest common factor of 42 and 15?

    c. What is the greatest common factor of 42 and 50?

NAME _____ DATE _____ PERIOD _____

3. A school chorus has 90 sixth-grade students and 75 seventh-grade students. The music director wants to make groups of performers, with the same combination of sixth- and seventh-grade students in each group. She wants to form as many groups as possible.

   a. What is the largest number of groups that could be formed? Explain or show your reasoning.

   b. If that many groups are formed, how many students of each grade level would be in each group?

4. Here are some bank transactions from a bank account last week. Which transactions represent negative values? Select **all** that apply. (Lesson 7-13)

   A. Monday: $650 paycheck deposited

   B. Tuesday: $40 withdrawal from the ATM at the gas pump

   C. Wednesday: $20 credit for returned merchandise

   D. Thursday: $125 deducted for cell phone charges

   E. Friday: $45 check written to pay for book order

   F. Saturday: $80 withdrawal for weekend spending money

   G. Sunday: $10 cash-back reward deposited from a credit card company

5. Find the quotients. (Lesson 4-11)

a. $\dfrac{1}{7} \div \dfrac{1}{8}$

b. $\dfrac{12}{5} \div \dfrac{6}{5}$

c. $\dfrac{1}{10} \div 10$

d. $\dfrac{9}{10} \div \dfrac{10}{9}$

6. An elephant can travel at a constant speed of 25 miles per hour, while a giraffe can travel at a constant speed of 16 miles in $\dfrac{1}{2}$ hour. (Lesson 2-9)

a. Which animal runs faster? Explain your reasoning.

b. How far can each animal run in 3 hours?

Lesson 7-17

Common Multiples

NAME _____ DATE _____ PERIOD _____

Learning Goal Let's use multiples to solve problems.

Warm Up
17.1 Notice and Wonder: Multiples

1. Circle all the multiples of 4 in this list.

   1 2 3 4 5 6 7 8 9 10 11 12 13 14 15 16 17 18 19 20 21 22 23 24 25 26

2. Circle all the multiples of 6 in this list.

   1 2 3 4 5 6 7 8 9 10 11 12 13 14 15 16 17 18 19 20 21 22 23 24 25 26

3. What do you notice? What do you wonder?

Activity
17.2 The Florist's Order

A florist can order roses in bunches of 12 and lilies in bunches of 8.
Last month she ordered the same number of roses and lilies.

1. If she ordered no more than 100 of each kind of flower, how many bunches of each could she have ordered? Find all the possible combinations.

2. What is the smallest number of bunches of roses that she could have ordered? What about the smallest number of bunches of lilies? Explain your reasoning.

Activity

17.3 Least Common Multiple

The **least common multiple** of 6 and 8 is 24.

1. What do you think the term "least common multiple" means?

2. Find all of the **multiples** of 10 and 8 that are less than 100. Find the least common multiple of 10 and 8.

3. Find all of the multiples of 7 and 9 that are less than 100. Find the least common multiple of 7 and 9.

Are you ready for more?

1. What is the least common multiple of 10 and 20?

2. What is the least common multiple of 4 and 12?

3. In the previous two questions, one number is a multiple of the other. What do you notice about their least common multiple? Do you think this will always happen when one number is a multiple of the other? Explain your reasoning.

NAME _____ DATE _____ PERIOD _____

Activity

17.4 Prizes on Grand Opening Day

Lin's uncle is opening a bakery. On the bakery's grand opening day, he plans to give away prizes to the first 50 customers that enter the shop. Every fifth customer will get a free bagel. Every ninth customer will get a free blueberry muffin. Every 12th customer will get a free slice of carrot cake.

1. Diego is waiting in line and is the 23rd customer. He thinks that he should get farther back in line in order to get a prize. Is he right? If so, how far back should he go to get at least one prize? Explain your reasoning.

2. Jada is the 36th customer.

   a. Will she get a prize? If so, what prize will she get?

   b. Is it possible for her to get more than one prize? How do you know? Explain your reasoning.

3. How many prizes total will Lin's uncle give away? Explain your reasoning.

Summary
Common Multiples

A multiple of a whole number is a product of that number with another whole number. For example, 20 is a multiple of 4 because $20 = 5 \cdot 4$.

A **common multiple** for two whole numbers is a number that is a multiple of both numbers. For example, 20 is a multiple of 2 and a multiple of 5, so 20 is a common multiple of 2 and 5.

The **least common multiple** (sometimes written as LCM) of two whole numbers is the smallest multiple they have in common. For example, 30 is the least common multiple of 6 and 10.

One way to find the least common multiple of two numbers is to list multiples of each in order until we find the smallest multiple they have in common. Let's find the least common multiple for 4 and 10. First, we list some multiples of each number.

- Multiples of 4: 4, 8, 12, 16, **20**, 24, 28, 32, 36, **40**, 44 . . .

- Multiples of 10: 10, **20**, 30, **40**, 50, . . .

20 and 40 are both common multiples of 4 and 10 (as are 60, 80, . . .), but 20 is the smallest number that is on *both* lists, so 20 is the least common multiple.

Glossary

common multiple
least common multiple

NAME _____ DATE _____ PERIOD _____

Practice
Common Multiples

1. Respond to each of the following.

   a. A green light blinks every 4 seconds and a yellow light blinks every 5 seconds. When will both lights blink at the same time?

   b. A red light blinks every 12 seconds and a blue light blinks every 9 seconds. When will both lights blink at the same time?

   c. Explain how to determine when 2 lights blink together.

2. Respond to each of the following.

   a. List all multiples of 10 up to 100.

   b. List all multiples of 15 up to 100.

   c. What is the least common multiple of 10 and 15?

3. Cups are sold in packages of 8. Napkins are sold in packages of 12.

   a. What is the fewest number of packages of cups and the fewest number of packages of napkins that can be purchased so there will be the same number of cups as napkins?

   b. How many sets of cups and napkins will there be?

4. Respond to each of the following. (Lesson 7-15)

   a. Plot and connect these points to form a polygon.

      (-5, 3), (3, 3), (1, -2), (-3, -2)

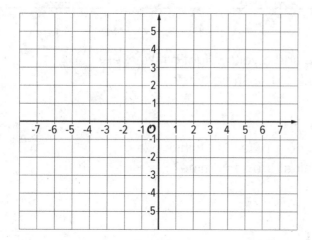

   b. Find the lengths of the two horizontal sides of the polygon.

5. Rectangle ABCD is drawn on a coordinate plane. A = (-6, 9) and B = (5, 9). (Lesson 7-14)

   What could be the locations of points C and D.

6. A school wants to raise $2,500 to support its music program. (Lesson 3-14)

   a. If it has met 20% of its goal so far, how much money has it raised?

   b. If it raises 175% of its goal, how much money will the music program receive? Show your reasoning.

Lesson 7-18

Using Common Multiples and Common Factors

NAME _____ DATE _____ PERIOD _____

Learning Goal Let's use common factors and common multiple to solve problems.

Warm Up
18.1 Keeping a Steady Beat

Your teacher will give you instructions for playing a rhythm game. As you play the game, think about these questions:

- When will the two sounds happen at the same time?

- How does this game relate to common factors or common multiples?

Activity
18.2 Factors and Multiples

Work with your partner to solve the following problems.

1. **Party.** Elena is buying cups and plates for her party. Cups are sold in packs of 8 and plates are sold in packs of 6. She wants to have the same number of plates and cups.

   a. Find a number of plates and cups that meets her requirement.

   b. How many packs of each supply will she need to buy to get that number?

   c. Name two other quantities of plates and cups she could get to meet her requirement.

2. **Tiles**. A restaurant owner is replacing the restaurant's bathroom floor with square tiles. The tiles will be laid side-by-side to cover the entire bathroom with no gaps, and none of the tiles can be cut. The floor is a rectangle that measures 24 feet by 18 feet.

   a. What is the largest possible tile size she could use? Write the side length in feet. Explain how you know it's the largest possible tile.

   b. How many of these largest size tiles are needed?

   c. Name more tile sizes that are whole number of feet that she could use to cover the bathroom floor. Write the side lengths (in feet) of the square tiles.

3. **Stickers**. To celebrate the first day of spring, Lin is putting stickers on some of the 100 lockers along one side of her middle school's hallway. She puts a skateboard sticker on every 4th locker (starting with locker 4), and a kite sticker on every 5th locker (starting with locker 5).

   a. Name three lockers that will get both stickers.

   b. After Lin makes her way down the hall, will the 30th locker have no stickers, 1 sticker, or 2 stickers? Explain how you know.

4. **Kits.** The school nurse is assembling first-aid kits for the teachers. She has 75 bandages and 90 throat lozenges. All the kits must have the same number of each supply, and all supplies must be used.

   a. What is the largest number of kits the nurse can make?

   b. How many bandages and lozenges will be in each kit?

NAME _____ DATE _____ PERIOD _____

5. What kind of mathematical work was involved in each of the previous problems? Put a checkmark to show what the questions were about.

Problem	Finding Multiples	Finding Least Common Multiples	Finding Factors	Finding Greatest Common Factor
Party				
Tiles				
Stickers				
Kits				

Are you ready for more?

You probably know how to draw a five-pointed star without lifting your pencil. One way to do this is to start with five dots arranged in a circle, then connect every second dot.

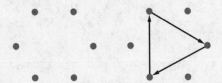

If you try the same thing with six dots arranged in a circle, you will have to lift your pencil. Once you make the first triangle, you'll have to find an empty dot and start the process over. Your six-pointed star has two pieces that are each drawn without lifting the pencil.

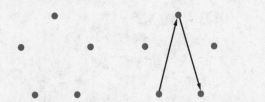

With twelve dots arranged in a circle, we can make some twelve-pointed stars.

1. Start with one dot and connect every second dot, as if you were drawing a five-pointed star. Can you draw the twelve-pointed star without lifting your pencil? If not, how many pieces does the twelve-pointed star have?

NAME _____ DATE _____ PERIOD _____

2. This time, connect every third dot. Can you draw this twelve-pointed star without lifting your pencil? If not, how many pieces do you get?

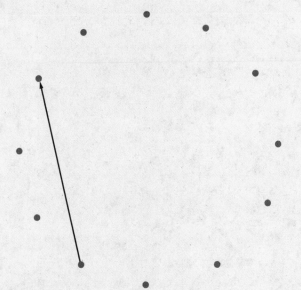

3. What do you think will happen if you connect every fourth dot? Try it. How many pieces do you get?

4. Do you think there is any way to draw a twelve-pointed star without lifting your pencil? Try it out.

5. Now investigate eight-pointed stars, nine-pointed stars, and ten-pointed stars. What patterns do you notice?

NAME _____ DATE _____ PERIOD _____

 Activity

18.3 More Factors and Multiples

Here are five more problems. Read and discuss each one with your group. *Without solving,* predict whether each problem involves finding common multiples or finding common factors. Circle one or more options to show your prediction.

1. **Soccer.** Diego and Andre are both in a summer soccer league. During the month of August, Diego has a game every 3rd day, starting August 3rd, and Andre has a game every 4th day, starting August 4th.

   - common multiples

   - least common multiple

   - common factors

   - greatest common factor

   a. What is the first date that both boys will have a game?

   b. How many of their games fall on the same date?

2. **Performances.** During a performing arts festival, students from elementary and middle schools will be grouped together for various performances. There are 32 elementary students and 40 middle-school students. The arts director wants identical groups for the performances, with students from both schools in each group. Each student can be a part of only one group.

   - common multiples

   - least common multiple

   - common factors

   - greatest common factor

   a. Name all possible groupings.

   b. What is the largest number of groups that can be formed? How many elementary school students and how many middle school students will be in each group?

3. **Lights.** There is a string of holiday lights with red, gold, and blue lights. The red lights are set to blink every 12 seconds, the gold lights are set to blink every 8 seconds, and the blue lights are set to blink every 6 seconds. The lights are on an automatic timer that starts each day at 7:00 p.m. and stops at midnight.

- common multiples

- least common multiple

- common factors

- greatest common factor

a. After how much time will all 3 lights blink at the exact same time?

b. How many times total will this happen in one day?

4. **Banners.** Noah has two pieces of cloth. He is making square banners for students to hold during the opening day game. One piece of cloth is 72 inches wide. The other is 90 inches wide. He wants to use all the cloth, and each square banner must be of equal width and as wide as possible.

- common multiples

- least common multiple

- common factors

- greatest common factor

a. How wide should he cut the banners?

b. How many banners can he cut?

NAME _____ DATE _____ PERIOD _____

5. **Dancers.** At Elena's dance recital her performance begins with a line of 48 dancers that perform in the dark with a black light that illuminates white clothing. All 48 dancers enter the stage in a straight line. Every 3rd dancer wears a white headband, every 5th dancer wears a white belt, and every 9th dancer wears a set of white gloves.

   • common multiples

   • least common multiple

   • common factors

   • greatest common factor

   a. If Elena is the 30th dancer, what accessories will she wear?

   b. Will any of the dancers wear all 3 accessories? If so, which one(s)?

   c. How many of each accessory will the dance teacher need to order?

6. Your teacher will assign your group a problem. Work with your group to solve the problem. Show your reasoning. Pause here so your teacher can review your work.

7. Work with your group to create a visual display that includes a diagram, an equation, and a math vocabulary word that would help to explain your mathematical thinking while solving the problem.

8. Prepare a short presentation in which all group members are involved. Your presentation should include: the problem (read aloud), your group's prediction of what mathematical concept the problem involved, and an explanation of each step of the solving process.

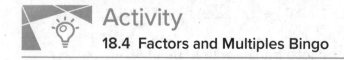

Activity

18.4 Factors and Multiples Bingo

Your teacher will explain the directions for a bingo game. Here are some things to keep in mind:

- Share one bingo board and some bingo chips with a partner.

- To play the game, your teacher will read statements aloud. You may help one another decide what numbers fit each statement, but speak only in a whisper. If the teacher hears anything above a whisper, you are out.

- The first person to call bingo needs to call out each number and identify the statement that it corresponds to. If there is an error in identifying statements, that player is out and the round continues.

Good luck, and have fun!

Summary

Using Common Multiples and Common Factors

If a problem requires dividing two whole numbers by the same whole number, solving it involves looking for a common factor. If it requires finding the *largest* number that can divide into the two whole numbers, we are looking for the *greatest common factor*.

Suppose we have 12 bagels and 18 muffins and want to make bags so each bag has the same combination of bagels and muffins. The common factors of 12 and 18 tell us possible number of bags that can be made.

The common factors of 18 are 1, 2, 3, and 6. For these numbers of bags, here are the number of bagels and muffins per bag.

- 1 bag: 12 bagels and 18 muffins

- 2 bags: 6 bagels and 9 muffins

- 3 bags: 4 bagels and 6 muffins

- 6 bags: 2 bagels and 3 muffins

We can see that the largest number of bags that can be made, 6, is the greatest common factor.

NAME _____ DATE _____ PERIOD _____

If a problem requires finding a number that is a multiple of two given numbers, solving it involves looking for a common multiple. If it requires finding the *first* instance the two numbers share a multiple, we are looking for the *least common multiple.*

Suppose forks are sold in boxes of 9 and spoons are sold in boxes of 15, and we want to buy an equal number of each. The multiples of 9 tell us how many forks we could buy, and the multiples of 15 tell us how many spoons we could buy, as shown here.

- Forks: 9, 18, 27, 36, 45, 54, 63, 72, 90. . .

- Spoons: 15, 30, 45, 60, 75, 90. . .

If we want as many forks as spoons, our options are 45, 90, 135, and so on, but the smallest number of utensils we could buy is 45, the least common multiple. This means buying 5 boxes of forks ($5 \cdot 9 = 45$) and 3 boxes of spoons ($3 \cdot 15 = 45$).

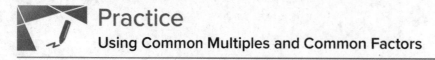

Practice

Using Common Multiples and Common Factors

1. Mai, Clare, and Noah are making signs to advertise the school dance. It takes Mai 6 minutes to complete a sign, it takes Clare 8 minutes to complete a sign, and it takes Noah 5 minutes to complete a sign. They keep working at the same rate for a half hour.

   a. Will Mai and Clare complete a sign at the same time? Explain your reasoning.

   b. Will Mai and Noah complete a sign at the same time? Explain your reasoning.

   c. Will Clare and Noah complete a sign at the same time? Explain your reasoning.

   d. Will all three students complete a sign at the same time? Explain your reasoning.

NAME _____ DATE _____ PERIOD _____

2. Diego has 48 chocolate chip cookies, 64 vanilla cookies, and 100 raisin cookies for a bake sale. He wants to make bags that have all three cookie flavors and the same number of each flavor per bag. (Lesson 7-16)

   a. How many bags can he make without having any cookies left over?

   b. Find the other solution to this problem.

3. Respond to each of the following.

   a. Find the product of 12 and 8.

   b. Find the greatest common factor of 12 and 8.

   c. Find the least common multiple of 12 and 8.

   d. Find the product of the greatest common factor and the least common multiple of 12 and 8.

   e. What do you notice about the answers to question 1 and question 4?

   f. Choose 2 other numbers and repeat the previous steps. Do you get the same results?

4. Respond to each of the following. (Lesson 7-11)

   a. Given the point (5.5, -7):

     i. Name a second point so that the two points form a vertical segment.

     ii. What is the length of the segment?

   b. Given the point (3, 3.5):

     i. Name a second point so that the two points form a horizontal segment.

     ii. What is the length of the segment?

5. Find the value of each expression mentally. (Lesson 6-9)

   a. $\frac{1}{2} \cdot 37 - \frac{1}{2} \cdot 7$

   b. $3.5 \cdot 40 + 3.5 \cdot 60$

   c. $999 \cdot 5$

Lesson 7-19

Drawing on the Coordinate Plane

NAME _____ DATE _____ PERIOD _____

Learning Goal Let's draw on the coordinate plane.

 ## Activity

19.1 Cat Pictures

Use graphing technology to recreate this image. If graphing technology is not available, list the ordered pairs that make up this image. Then compare your list with a partner.

If you have time, consider adding more details to your image such as whiskers, the inside of the ears, a bow, or a body.

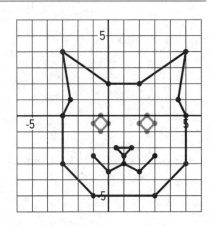

Are you ready for more?

If you are using graphing technology, add these statements to the list of things being graphed:

$x > 6$

$y > 5$

$x < -4$

$y < -6$

Describe the result. Why do you think that happened?

Activity

19.2 Design Your Own Image

Use graphing technology to create an image of your own design. You could draw a different animal, a vehicle, a building, or something else. Make sure your image includes at least 4 points in each quadrant of the coordinate plane.

If graphing technology is not available, create your image on graph paper, and then list the ordered pairs that make up your image. Trade lists with a partner but do not show them your image. Graph your partner's ordered pairs and see if your images match.

Learning Targets

Lesson	Learning Target(s)
7-1 Positive and Negative Numbers	• I can explain what 0, positive numbers, and negative numbers mean in the context of temperature and elevation.
	• I can use positive and negative numbers to describe temperature and elevation.
	• I know what positive and negative numbers are.
7-2 Points on the Number Line	• I can determine or approximate the value of any point on a number line.
	• I can represent negative numbers on a number line.
	• I understand what it means for numbers to be opposites.
7-3 Comparing Positive and Negative Numbers	• I can explain how to use the positions of numbers on a number line to compare them.
	• I can explain what a rational number is.
	• I can use inequalities to compare positive and negative numbers.

(continued on the next page)

(continued from the previous page)

Lesson	Learning Target(s)
7-4 Ordering Rational Numbers	• I can compare and order rational numbers. • I can use phrases like "greater than," "less than," and "opposite" to compare rational numbers.
7-5 Using Negative Numbers to Make Sense of Contexts	• I can explain and use negative numbers in situations involving money. • I can interpret and use negative numbers in different contexts.
7-6 Absolute Value of Numbers	• I can explain what the absolute value of a number is. • I can find the absolute values of rational numbers. • I can recognize and use the notation for absolute value.
7-7 Comparing Numbers and Distance from Zero	• I can explain what absolute value means in situations involving elevation. • I can use absolute values to describe elevations. • I can use inequalities to compare rational numbers and the absolute values of rational numbers.

Lesson	Learning Target(s)
7-8 Writing and Graphing Inequalities	• I can graph inequalities on a number line. • I can write an inequality to represent a situation.
7-9 Solutions of Inequalities	• I can determine if a particular number is a solution to an inequality. • I can explain what it means for a number to be a solution to an inequality. • I can graph the solutions to an inequality on a number line.
7-10 Interpreting Inequalities	• I can explain what the solution to an inequality means in a situation. • I can write inequalities that involve more than one variable.
7-11 Points on the Coordinate Plane	• I can describe a coordinate plane that has four quadrants. • I can plot points with negative coordinates in the coordinate plane. • I know what negative numbers in coordinates tell us.

(continued on the next page)

(continued from the previous page)

Lesson	Learning Target(s)
7-12 Constructing the Coordinate Plane	• When given points to plot, I can construct a coordinate plane with an appropriate scale and pair of axes.
7-13 Interpreting Points on a Coordinate Plane	• I can explain how rational numbers represent balances in a money context. • I can explain what points in a four-quadrant coordinate plane represent in a situation. • I can plot points in a four-quadrant coordinate plane to represent situations and solve problems.
7-14 Distances on a Coordinate Plane	• I can find horizontal and vertical distances between points on the coordinate plane.
7-15 Shapes on the Coordinate Plane	• I can find the lengths of horizontal and vertical segments in the coordinate plane. • I can plot polygons on the coordinate plane when I have the coordinates for the vertices.

Lesson	Learning Target(s)
7-16 Common Factors	• I can explain what a common factor is. • I can explain what the greatest common factor is. • I can find the greatest common factor of two whole numbers.
7-17 Common Multiples	• I can explain what a common multiple is. • I can explain what the least common multiple is. • I can find the least common multiple of two whole numbers.
7-18 Using Common Multiples and Common Factors	• I can solve problems using common factors and multiples.
7-19 Drawing on the Coordinate Plane	• I can use ordered pairs to draw a picture.

(continued on the next page)

Notes:

Data Sets and Distributions

Math, Data, and Bears, Oh My! In one of the upcoming lessons, you'll compare two sets of data that were collected by researchers studying wild bears.

Topics

- Data, Variability, and Statistical Questions
- Dot Plots and Histograms
- Mean and MAD
- Median and IQR
- Let's Put It to Work

Unit 8

Data Sets and Distributions

Lesson 8-1

Got Data?

NAME _____ DATE _____ PERIOD _____

Learning Goal Let's explore different kinds of data.

 **Warm Up**
1.1 Dots of Data

Here is a **dot plot** for a data set.

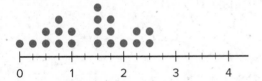

1. Determine if each of the following would be an appropriate label to represent the data in the dot plot. Be prepared to explain your reasoning.

   a. Number of children per class

   b. Distance between home and school, in miles

   c. Hours spent watching TV each day

   d. Weight of elephants, in pounds

   e. Points received on a homework assignment

2. Think of another label that can be used with the dot plot.

   a. Write it below the scale of the dot plot. Be sure to include the unit of measurement.

   b. In your scenario, what does one dot represent?

   c. In your scenario, what would a data point of 0 mean? What would a data point of $3\frac{1}{4}$ mean?

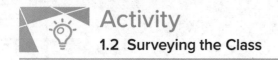

Activity

1.2 Surveying the Class

Here are some survey questions. Your teacher will explain which questions can be used to learn more about the students in your class and how the responses will be collected. The data that your class collects will be used in upcoming activities.

1. How long does it usually take you to travel to school? Answer to the nearest minute.

2. How do you travel to school on most days? Choose one.

   - Walk
   - Bike
   - Scooter or skateboard
   - Car

   - School bus
   - Public transport
   - Other

3. How tall are you without your shoes on? Answer to the nearest centimeter.

4. What is the length of your right foot without your shoe on? Answer to the nearest centimeter.

5. What is your arm span? Stretch your arms open, and measure the distance from the tip of your right hand's middle finger to the tip of your left hand's middle finger, across your back. Answer to the nearest centimeter.

6. How important are the following issues to you? Rate each on a scale from 0 (not important) to 10 (very important).

   a. Reducing pollution

   b. Recycling

   c. Conserving water

7. Do you have any siblings? _____ Yes _____ No

NAME _____ DATE _____ PERIOD _____

8. How many hours of sleep per night do you usually get when you have school the next day? Answer to the nearest half hour.

9. How many hours of sleep per night do you usually get when you do not have school the next day? Answer to the nearest half hour.

10. Other than traveling from school, what do you do right after school on most days?

- Have a snack.
- Do homework.
- Read a book.
- Talk on the phone.

- Practice a sport.
- Do chores.
- Use the computer.
- Participate in an extracurricular activity.

11. If you could meet one of these celebrities, who would you choose?

- A city or state leader
- A champion athlete
- A movie star

- A musical artist
- A best-selling author

12. Estimate how much time per week you usually spend on each of these activities. Answer to the nearest quarter of an hour.

a. Playing sports or doing outdoor activities

b. Using a screen for fun (watching TV, playing computer games, etc.)

c. Doing homework

d. Reading

Activity

1.3 Numerical and Categorical Data

The list of survey questions in the activity earlier can help you complete these exercises.

1. The first survey question about travel *time* produces **numerical data**. Identify two other questions that produce numerical data. For each, describe what was measured and its unit of measurement.

   a. Question #: _____   What was measured:

                                    Unit of measurement:

   b. Question #: _____   What was measured:

                                    Unit of measurement:

2. The second survey question about travel *method* produces **categorical data**. Identify two other questions that produce categorical data. For each, describe what characteristic or feature was being studied.

   a. Question #: _____   Characteristic being studied:

   b. Question #: _____   Characteristic being studied:

3. Think about the responses to these survey questions. Do they produce numerical or categorical data? Be prepared to explain how you know.

   a. How many pets do you have?   b. How many years have you lived in this state?

   c. What is your favorite band?   d. What kind of music do you like best?

   e. What is the area code of your school's phone number?   f. Where were you born?

   g. How much does your backpack weigh?

NAME _____ DATE _____ PERIOD _____

4. Name two characteristics you could investigate to learn more about your classmates. Make sure one would give categorical data and the other would give numerical data.

Are you ready for more?

Priya and Han collected data on the birth months of students in their class. Here are the lists of their records for the same group of students.

This list shows Priya's records.

Jan	Apr	Jan	Feb	Oct	May	June	July
Aug	Aug	Sep	Jan	Feb	Mar	Apr	Nov
Nov	Dec	Feb	Mar				

This list shows Han's records.

1	4	1	2	10	5	6	7
8	8	9	1	2	3	4	11
11	12	2	3				

1. How are their records alike? How are they different?

2. What kind of data—categorical or numerical—do you think the variable "birth month" produces? Explain how you know.

Summary
Got Data?

The table contains data about 10 dogs.

Dog Name	Weight (kg)	Breed
Duke	36	German Shepherd
Coco	6	Pug
Pierre	7	Pug
Ginger	35	German Shepherd
Lucky	10	Beagle
Daisy	10	Beagle
Buster	35	German Shepherd
Pepper	7	Pug
Rocky	7	Beagle
Lady	32	German Shepherd

- The weights of the dogs are an example of **numerical data**, which is data that are numbers, quantities, or measurements. The weights of the dogs are measurements in kilograms.

- The dog breeds are an example of **categorical data**, which is data containing values that can be sorted into categories. In this case, there are three categories for dog breeds: pug, beagle, and German shepherd.

Some data with numbers are categorical because the numbers are *not* quantities or measurements. For example, telephone area codes are categorical data, because the numbers are labels rather than quantities or measurements.

NAME _____ DATE _____ PERIOD _____

Numerical data can be represented with a **dot plot** (sometimes called a line plot). Here is a dot plot that shows the weights of the dogs.

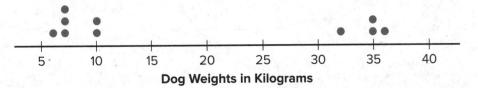

Dog Weights in Kilograms

We can collect and study both kinds of data by doing surveys or taking measurements. When we do, it is important to think about what feature we are studying (for example, breeds of dogs or weights of dogs) and what units of measurement are used.

Glossary

categorical data
dot plot
numerical data

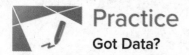
1. Tyler asked 10 students at his school how much time in minutes it takes them to get from home to school. Determine if each of these dot plots could represent the data Tyler collected. Explain your reasoning for each dot plot.

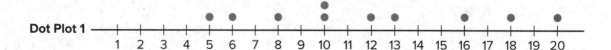

Dot Plot 1

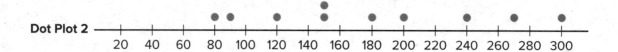

Dot Plot 2

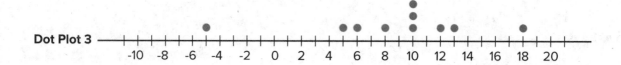

Dot Plot 3

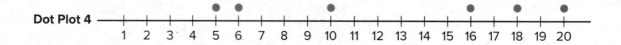

Dot Plot 4

NAME _____ DATE _____ PERIOD _____

2. Here is a list of questions. For each question, decide if the responses will produce numerical data or categorical data and give two possible responses.

a. What is your favorite breakfast food?

b. How did you get to school this morning?

c. How many different teachers do you have?

d. What is the last thing you ate or drank?

e. How many minutes did it take you to get ready this morning—from waking up to leaving for school?

3. Respond to each of the following.

a. Write two questions that you could ask the students in your class that would result in categorical data. For each question, explain how you know that responses to it would produce categorical data.

b. Write two questions that you could ask the students in your class that would result in numerical data. For each question, explain how you know that responses to it would produce numerical data.

3. Triangle *DEF* has vertices $D = (-4, -4)$, $E = (-2, -4)$, and $F = (-3, -1)$. (Lesson 7-15)

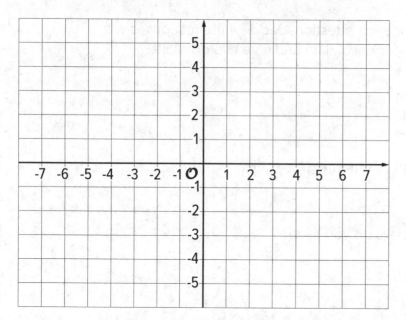

a. Plot the triangle in the coordinate plane and label the vertices.

b. Name the coordinates of 3 points that are inside the triangle.

c. What is the area of the triangle? Show your reasoning.

Lesson 8-2

Statistical Questions

NAME _____ DATE _____ PERIOD _____

Learning Goal Let's look more closely at data and the questions they can help to answer.

Warm Up
2.1 Pencils on a Plot

1. Measure your pencil to the nearest $\frac{1}{4}$-inch. Then, plot your measurement on the class dot plot.

2. What is the difference between the longest and shortest pencil lengths in the class?

3. What is the most common pencil length?

4. Find the difference in lengths between the most common length and the shortest pencil.

Activity

2.2 What's in the Data?

Ten sixth-grade students at a school were each asked five survey questions. Their answers to each question are shown here.

Data Set A	0	1	1	3	0	0	0	2	1	1
Data Set B	12	12	12	12	12	12	12	12	12	12
Data Set C	6	5	7	6	4	5	3	4	6	8
Data Set D	6	6	6	6	6	6	6	6	6	6
Data Set E	3	7	9	11	6	4	2	16	6	10

1. Here are the five survey questions. Match each question to a data set that could represent the students' answers. Explain your reasoning.

   - Question 1: Flip a coin 10 times. How many heads did you get?

   - Question 2: How many books did you read in the last year?

   - Question 3: What grade are you in?

   - Question 4: How many dogs and cats do you have?

   - Question 5: How many inches are in 1 foot?

2. How are survey questions 3 and 5 different from the other questions?

Activity

2.3 What Makes a Statistical Question?

These three questions are examples of **statistical questions**:

- What is the most common color of the cars in the school parking lot?

- What percentage of students in the school have a cell phone?

- Which kind of literature—fiction or nonfiction—is more popular among students in the school?

These three questions are not examples of statistical questions:

- What color is the principal's car?

- Does Elena have a cell phone?

- What kind of literature—fiction or nonfiction—does Diego prefer?

NAME _____ DATE _____ PERIOD _____

1. Study the examples and non-examples. Discuss with your partner:

   a. How are the three statistical questions alike? What do they have in common?

   b. How are the three non-statistical questions alike? What do they have in common?

   c. How can you find answers to the statistical questions? How about answers to non-statistical questions?

   d. What makes a question a statistical question?

Pause here for a class discussion.

2. Read each question. Think about the data you might collect to answer it and whether you expect to see **variability** in the data. Complete each blank with "Yes" or "No."

   a. How many cups of water do my classmates drink each day?

      • Is variability expected in the data? _____

      • Is the question statistical? _____

   b. Where in town does our math teacher live?

      • Is variability expected in the data? _____

      • Is the question statistical? _____

   c. How many minutes does it take students in my class to get ready for school in the morning?

      • Is variability expected in the data? _____

      • Is the question statistical? _____

   d. How many minutes of recess do sixth-grade students have each day?

      • Is variability expected in the data? _____

      • Is the question statistical? _____

   e. Do all students in my class know what month it is?

      • Is variability expected in the data? _____

      • Is the question statistical? _____

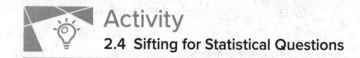

Activity

2.4 Sifting for Statistical Questions

1. Your teacher will give you and your partner a set of cards with questions. Sort them into three piles: Statistical Questions, Not Statistical Questions, and Unsure.

2. Compare your sorting decisions with another group's decisions. Start by discussing the two piles that your group sorted into the Statistical Questions and Not Statistical Questions piles. Then, review the cards in the Unsure pile. Discuss the questions until both groups reach an agreement and have no cards left in the Unsure pile. If you get stuck, think about whether the question could be answered by collecting data and if there would be variability in that data.

3. Record the letter names of the questions in each pile.

   - Statistical questions:

   - Non-statistical questions:

Are you ready for more?

Tyler and Han are discussing the question, "Which sixth-grade student lives the farthest from school?"

- Tyler says, "I don't think the question is a statistical question. There is only one person who lives the farthest from school, so there would not be variability in the data we collect."

- Han says: "I think it is a statistical question. To answer the question, we wouldn't actually be asking everyone, 'Which student lives the farthest from school?' We would have to ask each student how far away from school they live, and we can expect their responses to have variability."

Do you agree with either one of them? Explain your reasoning.

NAME _____ DATE _____ PERIOD _____

Summary
Statistical Questions

We often collect data to answer questions about something. The data we collect may show **variability**, which means the data values are not all the same.

Some data sets have more variability than others. Here are two sets of figures.

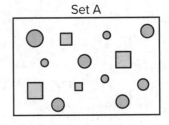
Set A

- Set A has more figures with the same shape, color, or size.

- Set B shows more figures with different shapes, colors, or sizes.

So, set B has greater variability than set A.

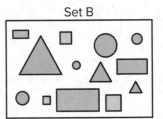
Set B

Both numerical and categorical data can show variability. Numerical sets can contain different numbers, and categorical sets can contain different categories or types.

When a question can only be answered by using data and we expect that data to have variability, we call it a **statistical question**. Here are some examples.

- Who is the most popular musical artist at your school?

- When do students in your class typically eat dinner?

- Which classroom in your school has the most books?

To answer the question about books, we may need to count all of the books in each classroom of a school. The data we collect would likely show variability because we would expect each classroom to have a different number of books.

In contrast, the question "How many books are in your classroom?" is *not* a statistical question. If we collect data to answer the question (for example, by asking everyone in the class to count books), the data can be expected to show the same value. Likewise, if we ask all of the students at a school where they go to school, that question is not a statistical question because the responses will all be the same.

Glossary

statistical question
variability

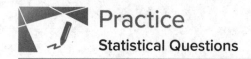

Practice
Statistical Questions

1. Sixth-grade students were asked, "What grade are you in?" Explain why this is *not* a statistical question.

2. Lin and her friends went out for ice cream after school. The following questions came up during their trip. Select **all** the questions that are statistical questions.

   (A.) How far are we from the ice cream shop?

   (B.) What is the most popular ice cream flavor this week?

   (C.) What does a group of 4 people typically spend on ice cream at this shop?

   (D.) Do kids usually prefer to get a cup or a cone?

   (E.) How many toppings are there to choose from?

3. Here is a list of questions about the students and teachers at a school. Select **all** the questions that are statistical questions.

   (A.) What is the most popular lunch choice?

   (B.) What school do these students attend?

   (C.) How many math teachers are in the school?

   (D.) What is a common age for the teachers at the school?

   (E.) About how many hours of sleep do students generally get on a school night?

   (F.) How do students usually travel from home to school?

NAME _____ DATE _____ PERIOD _____

4. Here is a list of statistical questions. What data would you collect and analyze to answer each question? For numerical data, include the unit of measurement that you would use.

   a. What is a typical height of female athletes on a team in the most recent international sporting event?

   b. Are most adults in the school football fans?

   c. How long do drivers generally need to wait at a red light in Washington, DC?

5. Describe the scale you would use on the coordinate plane to plot each set of points. What value would you assign to each unit of the grid? (Lesson 7-13)

   a. (1, -6), (-7, -8), (-3, 7), (0, 9)

   b. (-20, -30), (-40, 10), (20, -10), (5, -20)

   c. $\left(\frac{-1}{3}, -1\right), \left(\frac{2}{3}, -1\frac{1}{3}\right), \left(\frac{-4}{3}, \frac{2}{3}\right), \left(\frac{1}{6}, 0\right)$

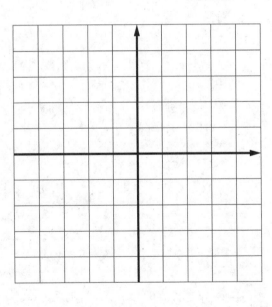

6. Noah's water bottle contains more than 1 quart of water but less than $1\frac{1}{2}$ quarts. Let w be the amount of water in Noah's bottle, in quarts. Select **all** the true statements. (Lesson 7-9)

(A.) w could be $\frac{3}{4}$.

(B.) w could be 1.

(C.) $w > 1$

(D.) w could be $\frac{4}{3}$.

(E.) w could be $\frac{5}{4}$.

(F.) w could be $\frac{5}{3}$.

(G.) $w > 1.5$

7. Order these numbers from least to greatest: (Lesson 7-7)

$|-17|$ $|-18|$ -18 $|19|$ 20

Lesson 8-3

Representing Data Graphically

NAME _____ DATE _____ PERIOD _____

Learning Goal Let's represent data with dot plots and bar graphs.

Warm Up
3.1 Curious about Caps

Clare collects bottle caps and keeps them in plastic containers. Write one statistical question that someone could ask Clare about her collection. Be prepared to explain your reasoning.

Activity
3.2 Estimating Caps

1. Write down the statistical question your class is trying to answer.

2. Look at the dot plot that shows the data from your class. Write down one thing you notice and one thing you wonder about the dot plot.

3. Use the dot plot to answer the statistical question. Be prepared to explain your reasoning.

Activity

3.3 Been There, Done That!

Priya wants to know if basketball players on a men's team and a women's team have had prior experience in international competitions. She gathered data on the number of times the players were on a team before 2016.

Men's Team

3 0 0 0 0 1 0 0 0 0 0 0

Women's Team

2 3 3 1 0 2 0 1 1 0 3 1

1. Did Priya collect categorical or numerical data?

2. Organize the information on the two basketball teams into these tables.

Men's Basketball Team Players

Number of Prior Competitions	Frequency (number)
0	
1	
2	
3	
4	

Women's Basketball Team Players

Number of Prior Competitions	Frequency (number)
0	
1	
2	
3	
4	

3. Make a dot plot for each table.

Men's Basketball Team Players

Women's Basketball Team Players

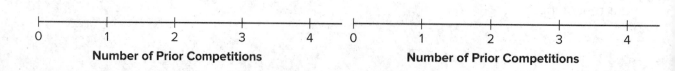

NAME _____ DATE _____ PERIOD _____

4. Study your dot plots. What do they tell you about the competition participation of:

    a. the players on the men's basketball team?

    b. the players on the women's basketball team?

5. Explain why a dot plot is an appropriate representation for Priya's data.

Are you ready for more?

Combine the data for the players on the men's and women's teams and represent it as a single dot plot. What can you say about the repeat participation of the basketball players?

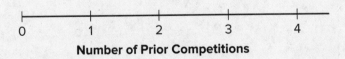

Number of Prior Competitions

Activity

3.4 Favorite Summer Sports

Kiran wants to know which three summer sports are most popular in his class. He surveyed his classmates on their favorite summer sport. Here are their responses.

swimming	basketball	volleyball
swimming	track and field	soccer
gymnastics	gymnastics	swimming
track and field	track and field	volleyball
diving	track and field	track and field
track and field	soccer	volleyball
gymnastics	track and field	gymnastics
diving	diving	rowing
gymnastics	basketball	swimming
swimming	volleyball	rowing

1. Did Kiran collect categorical or numerical data?

2. Organize the responses in a table to help him find which summer sports are most popular in his class.

Sport	Frequency

NAME _____ DATE _____ PERIOD _____

3. Represent the information in the table as a bar graph.

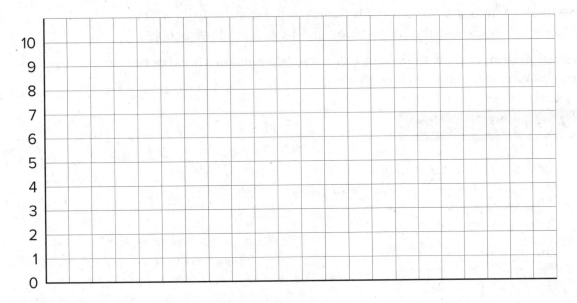

4. a. How can you use the bar graph to find how many classmates
      Kiran surveyed?

   b. Which three summer sports are most popular in Kiran's class?

   c. Use your bar graph to describe at least one observation about
      Kiran's classmates' preferred summer sports.

5. Could a dot plot be used to represent Kiran's data? Explain your reasoning.

When we analyze data, we are often interested in the **distribution**, which is information that shows all the data values and how often they occur.

In a previous lesson, we saw data about 10 dogs. We can see the distribution of the dog weights in a table such as this one.

Weight (kilograms)	Frequency
6	1
7	3
10	2
32	1
35	2
36	1

The term **frequency** refers to the number of times a data value occurs. In this case, we see that there are three dogs that weigh 7 kilograms, so "3" is the frequency for the value "7 kilograms."

Recall that dot plots are often used to represent numerical data. Like a frequency table, a dot plot also shows the distribution of a data set. This dot plot, which you saw in an earlier lesson, shows the distribution of dog weights.

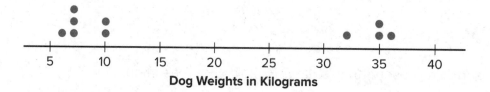

Dog Weights in Kilograms

A dot plot uses a horizontal number line. We show the frequency of a value by the number of dots drawn above that value. Here, the two dots above the number 35 tell us that there are two dogs weighing 35 kilograms.

NAME _____ DATE _____ PERIOD _____

The distribution of categorical data can also be shown in a table. This table shows the distribution of dog breeds.

Breed	Frequency
Pug	9
Beagle	9
German Shepherd	12

We often represent the distribution of categorical data using a bar graph.

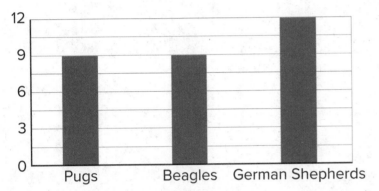

A bar graph also uses a horizontal line. Above it we draw a rectangle (or "bar") to represent each category in the data set. The height of a bar tells us the frequency of the category. There are 12 German shepherds in the data set, so the bar for this category is 12 units tall. Below the line we write the labels for the categories.

In a dot plot, a data value is placed according to its position on the number line. A weight of 10 kilograms must be shown as a dot above 10 on the number line.

In a bar graph, however, the categories can be listed in any order. The bar that shows the frequency of pugs can be placed anywhere along the horizontal line.

Glossary

distribution
frequency

1. A teacher drew a line segment that was 20 inches long on the blackboard. She asked each of her students to estimate the length of the segment and used their estimates to draw this dot plot.

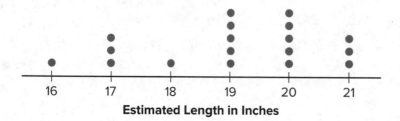

Estimated Length in Inches

   a. How many students were in the class?

   b. Were students generally accurate in their estimates of the length of the line? Explain your reasoning.

2. Here are descriptions of data sets. Select **all** the descriptions of data sets that could be graphed as dot plots.

   (A.) Class size for the classes at an elementary school

   (B.) Colors of cars in a parking lot

   (C.) Favorite sport of each student in a sixth-grade class

   (D.) Birth weights for the babies born during October at a hospital

   (E.) Number of goals scored in each of 20 games played by a school soccer team

NAME _____ DATE _____ PERIOD _____

3. Priya recorded the number of attempts it took each of 12 of her classmates to successfully throw a ball into a basket. Make a dot plot of Priya's data.

1   2   1   3   1   4   4   3   1   2   5   2

4. Solve each equation. (Lesson 6-4)

a. $9v = 1$

b. $1.37w = 0$

c. $1 = \frac{7}{10}x$

d. $12.1 = 12.1 + y$

e. $\frac{3}{5} + z = 1$

5. Find the quotients. (Lesson 4-11)

a. $\frac{2}{5} \div 2$

b. $\frac{2}{5} \div 5$

c. $2 \div \frac{2}{5}$

d. $5 \div \frac{2}{5}$

6. Find the area of each triangle. **(Lesson 1-9)**

a.

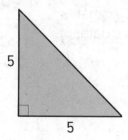

b.

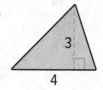

c.

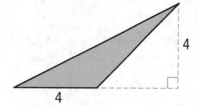

Lesson 8-4

Dot Plots

NAME _____ DATE _____ PERIOD _____

Learning Goal Let's investigate what dot plots and bar graphs can tell us.

Warm Up
4.1 Pizza Toppings (Part 1)

Fifteen customers in a pizza shop were asked, "How many toppings did you add to your cheese pizza?" Here are their responses.

1   2   1   3   0   1   1   2   0   3   0   0   1   2   2

1. Could you use a dot plot to represent the data? Explain your reasoning.

2. Complete the table.

Number of Toppings	Frequency (number)
0	
1	
2	
3	

Activity
4.2 Pizza Toppings (Part 2)

1. Use the tables from the warm-up to display the number of toppings as a dot plot. Label your drawing clearly.

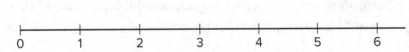

2. Use your dot plot to study the distribution for number of toppings. What do you notice about the number of toppings that this group of customers ordered? Write 2–3 sentences summarizing your observations.

Think of a statistical question that can be answered with the data about the number of toppings ordered, as displayed on the dot plot. Then answer this question.

Activity

4.3 Homework Time

Twenty-five sixth-grade students answered the question: "How many hours do you generally spend on homework each week?"

This dot plot shows the number of hours per week that these 25 students reported spending on homework.

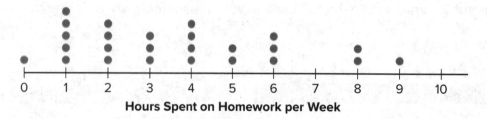

Hours Spent on Homework per Week

Use the dot plot to answer the following questions. For each, show or explain your reasoning.

1. What percentage of the students reported spending 1 hour on homework each week?

2. What percentage of the students reported spending 4 or fewer hours on homework each week?

3. Would 6 hours per week be a good description of the number of hours this group of students spends on homework per week? What about 1 hour per week? Explain your reasoning.

4. What value do you think would be a good description of the homework time of the students in this group? Explain your reasoning.

5. Someone said, "In general, these students spend roughly the same number of hours doing homework." Do you agree? Explain your reasoning.

NAME _____ DATE _____ PERIOD _____

Summary
Dot Plots

We often collect and analyze data because we are interested in learning what is "typical," or what is common and can be expected in a group.

Sometimes it is easy to tell what a typical member of the group is. For example, we can say that a typical shape in this set is a large circle.

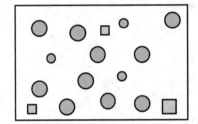

Just looking at the members of a group doesn't always tell us what is typical, however. For example, if we are interested in the side length typical of squares in this set, it isn't easy to do so just by studying the set visually.

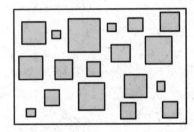

In a situation like this, it is helpful to gather the side lengths of the squares in the set and look at their distribution, as shown in this dot plot.

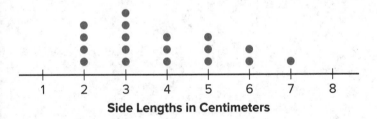

Side Lengths in Centimeters

We can see that many of the data points are between 2 and 4, so we could say that side lengths between 2 and 4 centimeters or close to these lengths are typical of squares in this set.

1. Clare recorded the amounts of time spent doing homework, in hours per week, by students in sixth, eighth, and tenth grades. She made a dot plot of the data for each grade and provided the following summary.

- Students in sixth grade tend to spend less time on homework than students in eighth and tenth grades.

- The homework times for the tenth-grade students are more alike than the homework times for the eighth-grade students.

Use Clare's summary to match each dot plot to the correct grade (sixth, eighth, or tenth).

a.

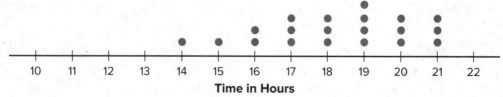

b.

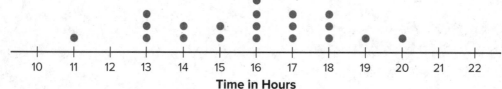

c.

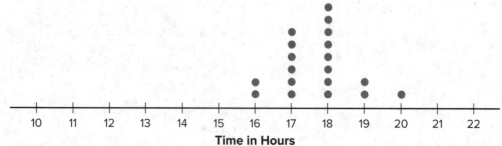

NAME _____ DATE _____ PERIOD _____

2. Mai played 10 basketball games. She recorded the number of points she scored and made a dot plot. Mai said that she scored between 8 and 14 points in most of the 10 games, but one game was exceptional. During that game she scored more than double her typical score of 9 points. Use the number line to make a dot plot that fits the description Mai gave.

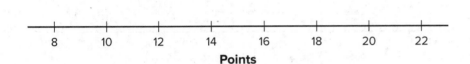

3. A movie theater is showing three different movies. The dot plots represent the ages of the people who were at the Saturday afternoon showing of each of these movies.

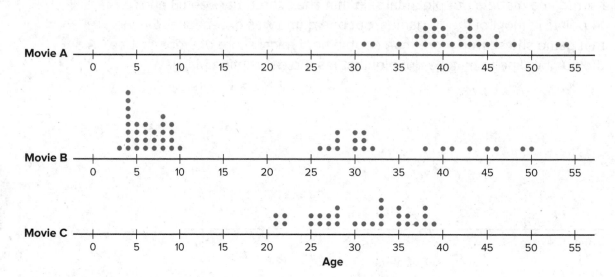

a. One of these movies was an animated movie rated G for general audiences. Do you think it was Movie A, B, or C? Explain your reasoning.

b. Which movie has a dot plot with ages that center at about 30 years?

c. What is a typical age for the people who were at Movie A?

4. Find the value of each expression. (Lesson 5-13)

a. $3.727 + 1.384$

b. $3.727 - 1.384$

c. $5.01 \cdot 4.8$

d. $5.01 \div 4.8$

Lesson 8-5

Using Dot Plots to Answer Statistical Questions

NAME _____ DATE _____ PERIOD _____

Learning Goal Let's use dot plots to describe distributions and answer questions.

Warm Up
5.1 Packs on Backs

This dot plot shows the weights of backpacks, in kilograms, of 50 sixth-grade students at a school in New Zealand.

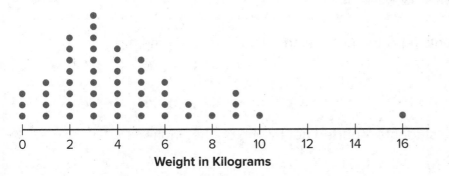

1. The dot plot shows several dots at 0 kilograms. What could a value of 0 mean in this context?

2. Clare and Tyler studied the dot plot.

   • Clare said, "I think we can use 3 kilograms to describe a typical backpack weight of the group because it represents 20%—or the largest portion—of the data."

   • Tyler disagreed and said, "I think 3 kilograms is too low to describe a typical weight. Half of the dots are for backpacks that are heavier than 3 kilograms, so I would use a larger value."

   Do you agree with either of them? Explain your reasoning.

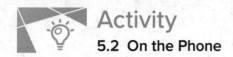

Activity

5.2 On the Phone

Twenty-five sixth-grade students were asked to estimate how many hours a week they spend talking on the phone. This dot plot represents their reported number of hours of phone usage per week.

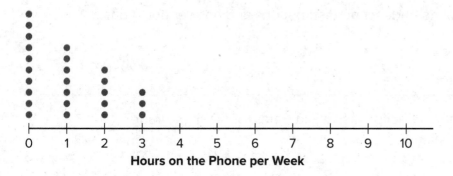

Hours on the Phone per Week

1. a. How many of the students reported not talking on the phone during the week? Explain how you know.

   b. What percentage of the students reported not talking on the phone?

2. a. What is the largest number of hours a student spent talking on the phone per week?

   b. What percentage of the group reported talking on the phone for this amount of time?

3. a. How many hours would you say that these students typically spend talking on the phone?

   b. How many minutes per day would that be?

NAME _____ DATE _____ PERIOD _____

4. a. How would you describe the **spread** of the data? Would you consider these students' amounts of time on the phone to be alike or different? Explain your reasoning.

b. Here is the dot plot from an earlier activity. It shows the number of hours per week the same group of 25 sixth-grade students reported spending on homework.

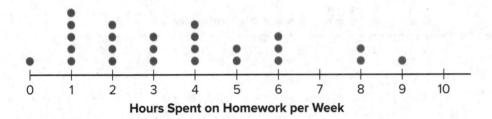

Hours Spent on Homework per Week

Overall, are these students more alike in the amount of time they spend talking on the phone or in the amount of time they spend on homework? Explain your reasoning.

5. Suppose someone claimed that these sixth-grade students spend too much time on the phone. Do you agree? Use your analysis of the dot plot to support your answer.

1. A keyboarding teacher wondered: "Do typing speeds of students improve after taking a keyboarding course?" Explain why her question is a statistical question.

2. The teacher recorded the number of words that her students could type per minute at the beginning of a course and again at the end. The two dot plots show the two data sets.

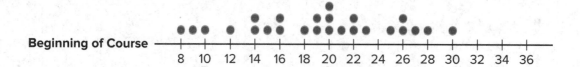

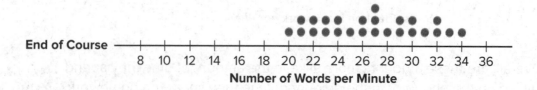

Based on the dot plots, do you agree with each of the following statements about this group of students? Be prepared to explain your reasoning.

   a. Overall, the students' typing speed did not improve. They typed at the same speed at the end of the course as they did at the beginning.

   b. 20 words per minute is a good estimate for how fast, in general, the students typed at the beginning of the course.

   c. 20 words per minute is a good description of the **center** of the data set at the end of the course.

   d. There was more variability in the typing speeds at the beginning of the course than at the end, so the students' typing speeds were more alike at the end.

3. Overall, how fast would you say that the students typed after completing the course? What would you consider the center of the end-of-course data?

NAME _____ DATE _____ PERIOD _____

Are you ready for more?

Use one of these suggestions (or make up your own). Research to create a dot plot with at least 10 values. Then, describe the center and spread of the distribution.

- Points scored by your favorite sports team in its last 10 games

- Length of your 10 favorite movies (in minutes)

- Ages of your favorite 10 celebrities

One way to describe what is typical or characteristic for a data set is by looking at the **center** and **spread** of its distribution.

Let's compare the distribution of cat weights and dog weights shown on these dot plots.

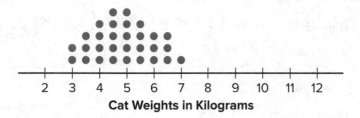

Cat Weights in Kilograms

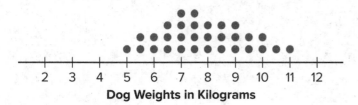

Dog Weights in Kilograms

The collection of points for the cat data is further to the left on the number line than the dog data. Based on the dot plots, we may describe the center of the distribution for cat weights to be between 4 and 5 kilograms and the center for dog weights to be between 7 and 8 kilograms.

We often say that values at or near the center of a distribution are typical for that group. This means that a weight of 4–5 kilograms is typical for a cat in the data set, and weight of 7–8 kilograms is typical for a dog.

NAME _____ DATE _____ PERIOD _____

We also see that the dog weights are more spread out than the cat weights. The difference between the heaviest and lightest cats is only 4 kilograms, but the difference between the heaviest and lightest dogs is 6 kilograms.

A distribution with greater spread tells us that the data have greater variability. In this case, we could say that the cats are more similar in their weights than the dogs.

In future lessons, we will discuss how to measure the center and spread of a distribution.

> **Glossary**
>
> **center**
> **spread**

Practice

Using Dot Plots to Answer Statistical Questions

1. Three sets of data about ten sixth-grade students were used to make three dot plots. The person who made these dot plots forgot to label them. Match each dot plot with the appropriate label.

Dot Plots

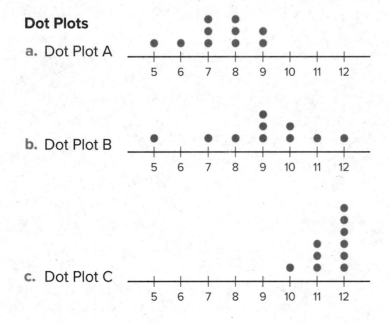

a. Dot Plot A

b. Dot Plot B

c. Dot Plot C

Labels

1. Ages in years

2. Numbers of hours of sleep on nights before school days

3. Numbers of hours of sleep on nights before non-school days

NAME _____ DATE _____ PERIOD _____

2. The dot plots show the time it takes to get to school for ten sixth-grade students from the United States, Canada, Australia, New Zealand, and South Africa.

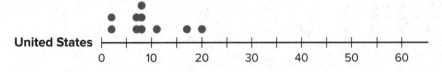

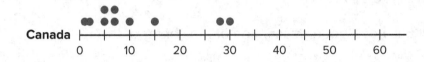

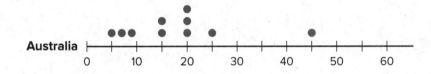

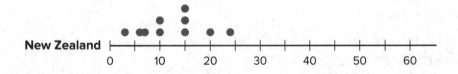

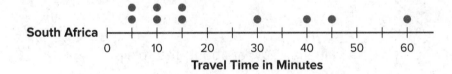

Travel Time in Minutes

a. List the countries in order of *typical travel times*, from shortest to longest.

b. List the countries in order of *variability in travel times*, from the least variability to the greatest.

3. Twenty-five students were asked to rate—on a scale of 0 to 10—how important it is to reduce pollution. A rating of 0 means "not at all important" and a rating of 10 means "very important." Here is a dot plot of their responses.

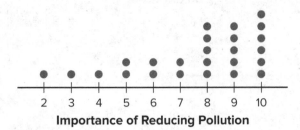

Importance of Reducing Pollution

Explain why a rating of 6 is *not* a good description of the center of this data set.

4. Tyler wants to buy some cherries at the farmer's market. He has $10 and cherries cost $4 per pound. (Lesson 7-10)

a. If c is the number of pounds of cherries that Tyler can buy, write one or more inequalities or equations describing c.

b. Can 2 be a value of c? Can 3 be a value of c? What about -1? Explain your reasoning.

c. If m is the amount of money, in dollars, Tyler can spend, write one or more inequalities or equations describing m.

d. Can 8 be a value of m? Can 2 be a value of m? What about 10.5? Explain your reasoning.

Lesson 8-6

Histograms

NAME _____ DATE _____ PERIOD _____

Learning Goal Let's explore how histograms represent data sets.

Warm Up
6.1 Dog Show (Part 1)

Here is a dot plot showing the weights, in pounds, of 40 dogs at a dog show.

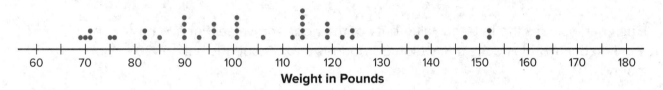

Weight in Pounds

1. Write two statistical questions that can be answered using the dot plot.

2. What would you consider a typical weight for a dog at this dog show?
   Explain your reasoning.

Activity

6.2 Dog Show (Part 2)

Here is a **histogram** that shows some dog weights in pounds.

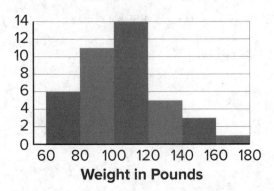

Each bar includes the left-end value but not the right-end value. For example, the first bar includes dogs that weigh 60 pounds and 68 pounds but not 80 pounds.

1. Use the histogram to answer the following questions.

   a. How many dogs weigh at least 100 pounds?

   b. How many dogs weigh exactly 70 pounds?

   c. How many dogs weigh at least 120 and less than 160 pounds?

   d. How much does the heaviest dog at the show weigh?

   e. What would you consider a typical weight for a dog at this dog show? Explain your reasoning.

2. Discuss with a partner:

   a. If you used the dot plot in Activity 6.1 to answer the same five questions you just answered, how would your answers be different?

   b. How are the histogram and the dot plot alike? How are they different?

NAME _____ DATE _____ PERIOD _____

Activity
6.3 Population of States

Every ten years, the United States conducts a census, which is an effort to count the entire population. The dot plot shows the population data from the 2010 census for each of the fifty states and the District of Columbia (DC).

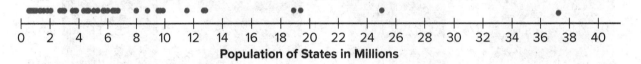

Population of States in Millions

1. Here are some statistical questions about the population of the fifty states and DC.

   How difficult would it be to answer the questions using the *dot plot*? In the middle column, rate each question with an E (easy to answer), H (hard to answer), or I (impossible to answer). Be prepared to explain your reasoning. (You will complete the *Histogram* column on the right side of the table in Exercise 4.)

Statistical Question	Using the Dot Plot	Using the Histogram
a. How many states have populations greater than 15 million?		
b. Which states have populations greater than 15 million?		
c. How many states have populations less than 5 million?		
d. What is a typical state population?		
e. Are there more states with fewer than 5 million people or more states with between 5 and 10 million people?		
f. How would you describe the distribution of state populations?		

2. Here are the population data for all states and the District of Columbia from the 2010 census. Use the information to complete the table.

Alabama	4.78	Illinois	12.83	Montana	0.99	Rhode Island	1.05
Alaska	0.71	Indiana	6.48	Nebraska	1.83	South Carolina	4.63
Arizona	6.39	Iowa	3.05	Nevada	2.70	South Dakota	0.81
Arkansas	2.92	Kansas	2.85	New Hampshire	1.32	Tennessee	6.35
California	37.25	Kentucky	4.34	New Jersey	8.79	Texas	25.15
Colorado	5.03	Louisiana	4.53	New Mexico	2.06	Utah	2.76
Connecticut	3.57	Maine	1.33	New York	19.38	Vermont	0.63
Delaware	0.90	Maryland	5.77	North Carolina	9.54	Virginia	8.00
District of Columbia	0.60	Massachusetts	6.55	North Dakota	0.67	Washington	6.72
Florida	18.80	Michigan	9.88	Ohio	11.54	West Virginia	1.85
Georgia	9.69	Minnesota	5.30	Oklahoma	3.75	Wisconsin	5.69
Hawaii	1.36	Mississippi	2.97	Oregon	3.83	Wyoming	0.56
Idaho	1.57	Missouri	5.99	Pennsylvania	12.70		

Population (millions)	Frequency
0–5	
5–10	
10–15	
15–20	
20–25	
25–30	
30–35	
35–40	

NAME _____ DATE _____ PERIOD _____

3. Use the grid and the information in your table to create a histogram.

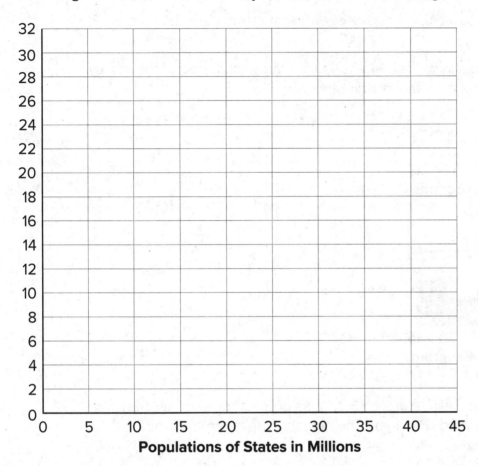

Populations of States in Millions

4. Return to the statistical questions at the beginning of the activity. Which ones are now easier to answer?
In the last column of the table, rate each question with an E (easy), H (hard), and I (impossible) based on how difficult it is to answer them. Be prepared to explain your reasoning.

Are you ready for more?

Think of two more statistical questions that can be answered using the data about populations of states. Then, decide whether each question can be answered using the dot plot, the histogram, or both.

In addition to using dot plots, we can also represent distributions of numerical data using **histograms**.

Here is a dot plot that shows the weights, in kilograms, of 30 dogs, followed by a histogram that shows the same distribution.

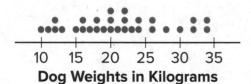

Dog Weights in Kilograms

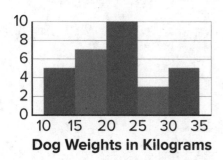

Dog Weights in Kilograms

In a histogram, data values are placed in groups or "bins" of a certain size, and each group is represented with a bar. The height of the bar tells us the frequency for that group.

For example, the height of the tallest bar is 10, and the bar represents weights from 20 to less than 25 kilograms, so there are 10 dogs whose weights fall in that group. Similarly, there are 3 dogs that weigh anywhere from 25 to less than 30 kilograms.

Notice that the histogram and the dot plot have a similar shape. The dot plot has the advantage of showing all of the data values, but the histogram is easier to draw and to interpret when there are a lot of values or when the values are all different.

NAME _____ DATE _____ PERIOD _____

Here is a dot plot showing the weight distribution of 40 dogs. The weights were measured to the nearest 0.1 kilogram instead of the nearest kilogram.

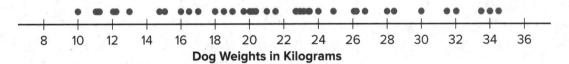

Dog Weights in Kilograms

Here is a histogram showing the same distribution.

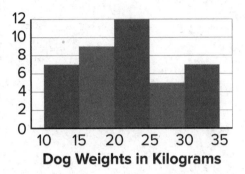

Dog Weights in Kilograms

In this case, it is difficult to make sense of the distribution from the dot plot because the dots are so close together and all in one line. The histogram of the same data set does a much better job showing the distribution of weights, even though we can't see the individual data values.

Glossary
histogram

1. Match histograms A through E to dot plots 1 through 5 so that each match represents the same data set.

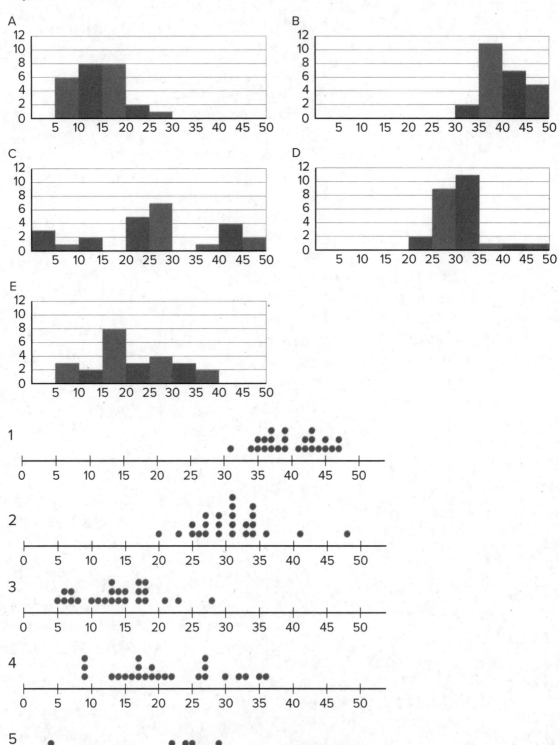

NAME _____ DATE _____ PERIOD _____

2. (-2, 3) is one vertex of a square on a coordinate plane. Name three points that could be the other vertices. **(Lesson 7-12)**

3. Here is a histogram that summarizes the lengths, in feet, of a group of adult female sharks. Select **all** the statements that are true, according to the histogram.

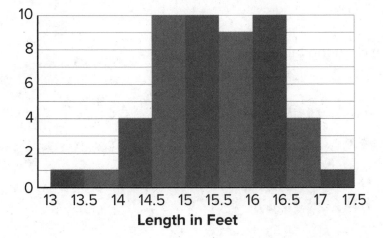

Length in Feet

(A.) A total of 9 sharks were measured.

(B.) A total of 50 sharks were measured.

(C.) The longest shark that was measured was 10 feet long.

(D.) Most of the sharks that were measured were over 16 feet long.

(E.) Two of the sharks that were measured were less than 14 feet long.

4. This table shows the times, in minutes, it took 40 sixth-grade students to run 1 mile.

Time (minutes)	Frequency
4 to less than 6	1
6 to less than 8	5
8 to less than 10	13
10 to less than 12	12
12 to less than 14	7
14 to less than 16	2

Draw a histogram for the information in the table.

Using Histograms to Answer Statistical Questions

NAME _____ DATE _____ PERIOD _____

Learning Goal Let's draw histograms and use them to answer questions.

Warm Up
7.1 Which One Doesn't Belong: Questions

Here are four questions about the population of Alaska. Which question does not belong? Be prepared to explain your reasoning.

- In general, at what age do Alaska residents retire?

- At what age can Alaskans vote?

- What is the age difference between the youngest and oldest Alaska residents with a full-time job?

- Which age group is the largest part of the population: 18 years or younger, 19–24 years, 25–34 years, 35–44 years, 45–54 years, 55–64 years, or 65 years or older?

An earthworm farmer set up several containers of a certain species of earthworms so that he could learn about their lengths. The lengths of the earthworms provide information about their ages. The farmer measured the lengths of 25 earthworms in one of the containers. Each length was measured in millimeters.

1. Using a ruler, draw a line segment for each length:

   - 20 millimeters

   - 40 millimeters

   - 60 millimeters

   - 80 millimeters

   - 100 millimeters

2. Here are the lengths, in millimeters, of the 25 earthworms.

6	11	18	19	20
23	23	25	25	26
27	27	28	29	32
33	41	42	48	52
54	59	60	77	93

   Complete the table for the lengths of the 25 earthworms.

Length	Frequency
0 millimeters to less than 20 millimeters	
20 millimeters to less than 40 millimeters	
40 millimeters to less than 60 millimeters	
60 millimeters to less than 80 millimeters	
80 millimeters to less than 100 millimeters	

NAME _____ DATE _____ PERIOD _____

3. Use the grid and the information in the table to draw a histogram for the worm length data. Be sure to label the axes of your histogram.

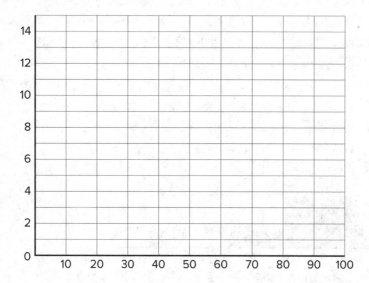

4. Based on the histogram, what is a typical length for these 25 earthworms? Explain how you know.

5. Write 1–2 sentences to describe the spread of the data. Do most of the worms have a length that is close to your estimate of a typical length, or are they very different in length?

Here is another histogram for the earthworm measurement data. In this histogram, the measurements are in different groupings.

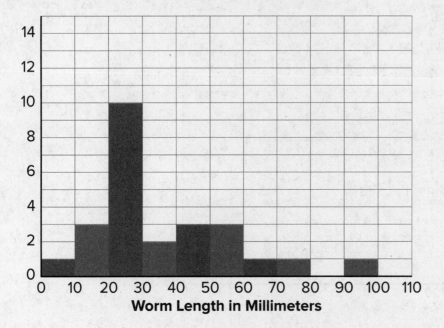

Worm Length in Millimeters

1. Based on this histogram, what is your estimate of a typical length for the 25 earthworms?

2. Compare this histogram with the one you drew. How are the distributions of data summarized in the two histograms the same? How are they different?

3. Compare your estimates of a typical earthworm length for the two histograms. Did you reach different conclusions about a typical earthworm length from the two histograms?

NAME _____ DATE _____ PERIOD _____

Activity

7.3 Tall and Taller Players

Professional basketball players tend to be taller than professional baseball players.

Here are two histograms that show height distributions of 50 male professional baseball players and 50 male professional basketball players.

1. Decide which histogram shows the heights of baseball players and which shows the heights of basketball players. Be prepared to explain your reasoning.

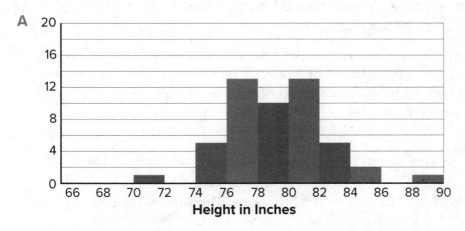

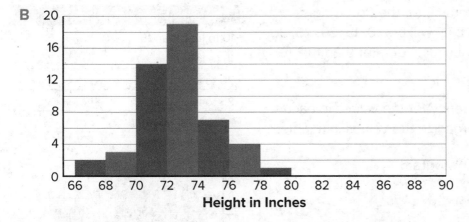

2. Write 2–3 sentences that describe the distribution of the heights of the basketball players. Comment on the center and spread of the data.

3. Write 2–3 sentences that describe the distribution of the heights of the baseball players. Comment on the center and spread of the data.

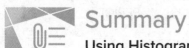

Summary

Using Histograms to Answer Statistical Questions

Here are the weights, in kilograms, of 30 dogs.

10	11	12	12	13	15	16	16	17	18
18	19	20	20	20	21	22	22	22	23
24	24	26	26	28	30	32	32	34	34

Before we draw a histogram, let's consider a couple of questions.

- What are the smallest and largest values in our data set? This gives us an idea of the distance on the number line that our histogram will cover. In this case, the minimum is 10 and the maximum is 34, so our number line needs to extend from 10 to 35 at the very least.

  (Remember the convention we use to mark off the number line for a histogram: we include the left boundary of a bar but exclude the right boundary. If 34 is the right boundary of the last bar, it won't be included in that bar, so the number line needs to go a little greater than the maximum value.)

- What group size or bin size seems reasonable here? We could organize the weights into bins of 2 kilograms (10, 12, 14, . . .), 5 kilograms, (10, 15, 20, 25, . . .), 10 kilograms (10, 20, 30, . . .), or any other size. The smaller the bins, the more bars we will have, and vice versa.

Let's use bins of 5 kilograms for the dog weights. The boundaries of our bins will be: 10, 15, 20, 25, 30, 35. We stop at 35 because it is greater than the maximum.

Weights (kilograms)	Frequency
10 to less than 15	5
15 to less than 20	7
20 to less than 25	10
25 to less than 30	3
30 to less than 35	5

Next, we find the frequency for the values in each group. It is helpful to organize the values in a table.

Now we can draw the histogram.

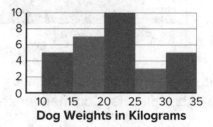

The histogram allows us to learn more about the dog weight distribution and describe its center and spread.

NAME _____ DATE _____ PERIOD _____

Practice
Using Histograms to Answer Statistical Questions

1. These two histograms show the number of text messages sent in one week by two groups of 100 students. The first histogram summarizes data from sixth-grade students. The second histogram summarizes data from seventh-grade students.

    a. Do the two data sets have approximately the same center? If so, explain where the center is located. If not, which one has the greater center?

    b. Which data set has greater spread? Explain your reasoning.

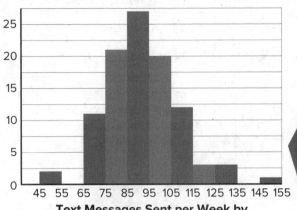

Text Messages Sent per Week by Sixth-Grade Students

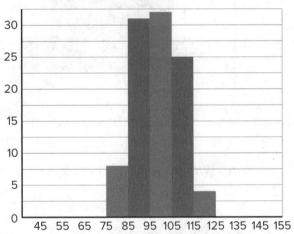

Text Messages Sent per Week by Seventh-Grade Students

   c. Overall, which group of students—sixth- or seventh-grade—sent more text messages?

2. Forty sixth-grade students ran 1 mile. Here is a histogram that summarizes their times, in minutes. The center of the distribution is approximately 10 minutes. (Lesson 7-9)

   On the blank axes, draw a second histogram that has:

   - a distribution of times for a different group of 40 sixth-grade students.

   - a center at 10 minutes.

   - less variability than the distribution shown in the first histogram.

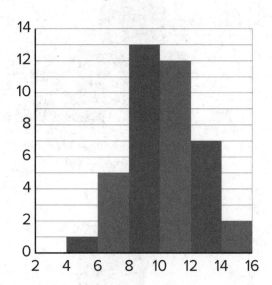 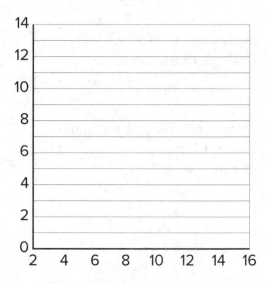

3. Jada has d dimes. She has more than 30 cents but less than a dollar. (Lesson 7-9)

   a. Write two inequalities that represent how many dimes Jada has.

   b. Can d be 10?

   c. How many possible solutions make both inequalities true? If possible, describe or list the solutions.

4. Order these numbers from greatest to least:

   $-4, \frac{1}{4}, 0, 4, -3\frac{1}{2}, \frac{7}{4}, -\frac{5}{4}$ (Lesson 7-4)

Lesson 8-8

Describing Distributions on Histograms

NAME _____ DATE _____ PERIOD _____

Learning Goal Let's describe distributions displayed in histograms.

Warm Up
8.1 Which One Doesn't Belong: Histograms

Which histogram does not belong? Be prepared to explain your reasoning.

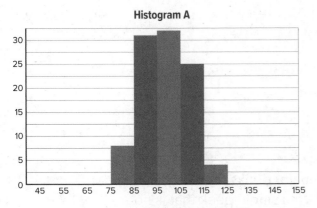

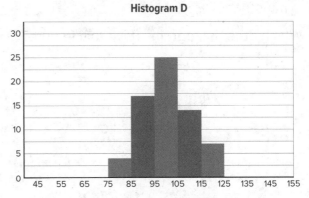

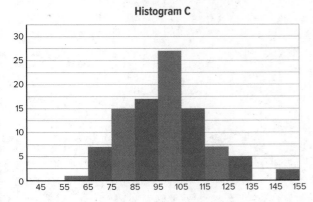

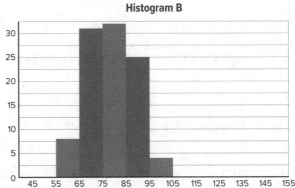

Activity

8.2 Sorting Histograms

1. Your teacher will give your group a set of histogram cards. Sort them into two piles—one for histograms that are approximately symmetrical, and another for those that are not.

2. Discuss your sorting decisions with another group. Do both groups agree which cards should go in each pile? If not, discuss the reasons behind the differences and see if you can reach agreement. Record your final decisions.

   • Histograms that are approximately symmetrical:

   • Histograms that are not approximately symmetrical:

3. Histograms are also described by how many major peaks they have. Histogram A is an example of a distribution with a single peak that is not symmetrical.

   Which other histograms have this feature?

NAME _____ DATE _____ PERIOD _____

4. Some histograms have a gap, a space between two bars where there are no data points. For example, if some students in a class have 7 or more siblings, but the rest of the students have 0, 1, or 2 siblings, the histogram for this data set would show gaps between the bars because no students have 3, 4, 5, or 6 siblings.

Which histograms do you think show one or more gaps?

5. Sometimes there are a few data points in a data set that are far from the center. Histogram A is an example of a distribution with this feature. Which other histograms have this feature?

Your teacher will provide you with some data that your class collected the other day.

1. Use the data to draw a histogram that shows your class's travel times.

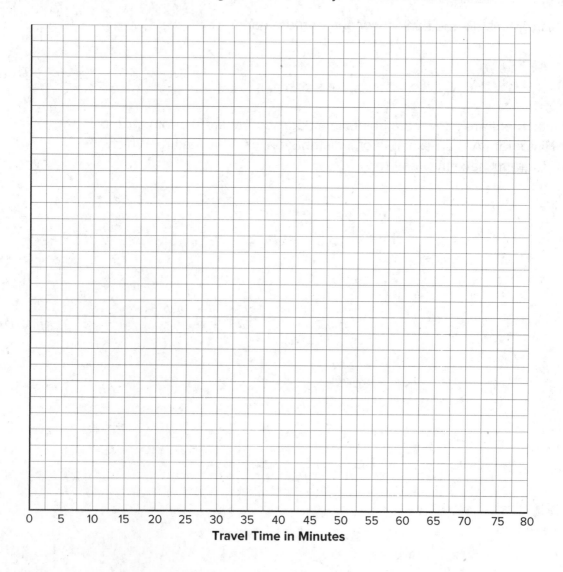

Travel Time in Minutes

2. Describe the distribution of travel times. Comment on the center and spread of the data, as well as the shape and features.

NAME _____ DATE _____ PERIOD _____

3. Use the data on methods of travel to draw a bar graph. Include labels for the horizontal axis.

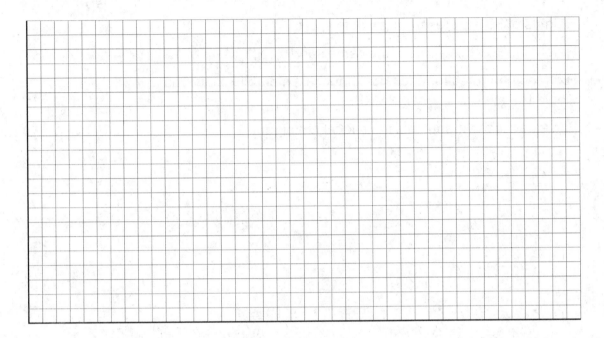

4. Describe what you learned about your class's methods of transportation to school. Comment on any patterns you noticed.

5. Compare the histogram and the bar graph that you drew.

How are they the same?

How are they different?

Use one of these suggestions (or make up your own). Research data to create a histogram. Then, describe the distribution.

- Heights of 30 athletes from multiple sports

- Heights of 30 athletes from the same sport

- High temperatures for each day of the last month in a city you would like to visit

- Prices for all the menu items at a local restaurant

NAME _____ DATE _____ PERIOD _____

Summary
Describing Distributions on Histograms

We can describe the shape and features of the distribution shown on a histogram. Here are two distributions with very different shapes and features.

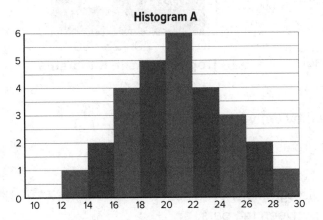

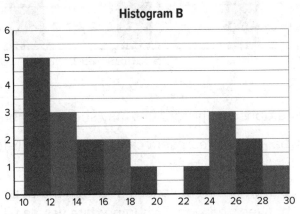

- Histogram A is very symmetrical and has a peak near 21. Histogram B is not symmetrical and has two peaks, one near 11 and one near 25.

- Histogram B has two clusters. A cluster forms when many data points are near a particular value (or a neighborhood of values) on a number line.

- Histogram B also has a gap between 20 and 22. A gap shows a location with no data values.

Here is a bar graph showing the breeds of 30 dogs and a histogram for their weights.

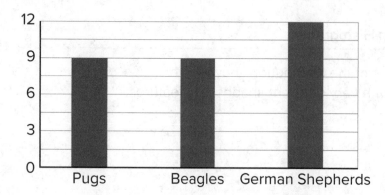

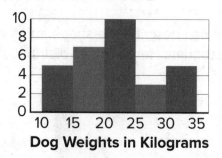

Dog Weights in Kilograms

Bar graphs and histograms may seem alike, but they are very different.

- Bar graphs represent categorical data. Histograms represent numerical data.

- Bar graphs have spaces between the bars. Histograms show a space between bars *only* when no data values fall between the bars.

- Bars in a bar graph can be in any order. Histograms must be in numerical order.

- In a bar graph, the number of bars depends on the number of categories. In a histogram, we choose how many bars to use.

NAME _____ DATE _____ PERIOD _____

Practice
Describing Distributions on Histograms

1. The histogram summarizes the data on the body lengths of 143 wild bears. Describe the distribution of body lengths. Be sure to comment on shape, center, and spread.

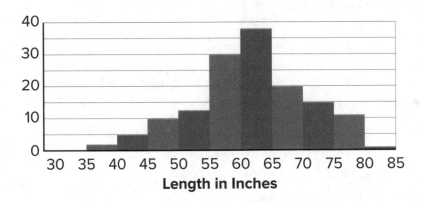

2. Which data set is more likely to produce a histogram with a symmetric distribution? Explain your reasoning.

   • Data on the number of seconds on a track of music in a pop album.

   • Data on the number of seconds spent talking on the phone yesterday by everyone in the school.

3. Evaluate the expression $4x^3$ for each value of x. (Lesson 6-15)

   a. 1   b. 2   c. $\frac{1}{2}$

4. Decide if each data set might produce one or more gaps when represented by a histogram. For each data set that you think might produce gaps, briefly describe or give an example of how the values in the data set might do so.

   a. The ages of students in a sixth-grade class

   b. The ages of people in an elementary school

   c. The ages of people eating in a family restaurant

   d. The ages of people who watch football

   e. The ages of runners in a marathon

5. Jada drank 12 ounces of water from her bottle. This is 60% of the water the bottle holds. (Lesson 6-7)

   a. Write an equation to represent this situation. Explain the meaning of any variables you use.

   b. How much water does the bottle hold?

Lesson 8-9

Interpreting the Mean as Fair Share

NAME _____ DATE _____ PERIOD _____

Learning Goal Let's explore the mean of a data set and what it tells us.

Warm Up
9.1 Close to Four

Use the digits 0–9 to write an expression with a value as close as possible to 4. Each digit can be used only one time in the expression.

$$\left(\boxed{} + \boxed{} + \boxed{} + \boxed{}\right) \div 4$$

Activity
9.2 Spread Out and Share

1. The kittens in a room at an animal shelter are placed in 5 crates.

a. The manager of the shelter wants the kittens distributed equally among the crates. How might that be done? How many kittens will end up in each crate?

b. The number of kittens in each crate after they are equally distributed is called the **mean** number of kittens per crate, or the **average** number of kittens per crate. Explain how the expression 10 ÷ 5 is related to the average.

c. Another room in the shelter has 6 crates. No two crates have the same number of kittens, and there is an average of 3 kittens per crate. Draw or describe at least two different arrangements of kittens that match this description.

2. Five servers were scheduled to work the number of hours shown. They decided to share the workload, so each one would work equal hours.

Server A: 3   Server B: 6   Server C: 11   Server D: 7   Server E: 4

a. On the first grid, draw 5 bars whose heights represent the hours worked by servers A, B, C, D, and E.

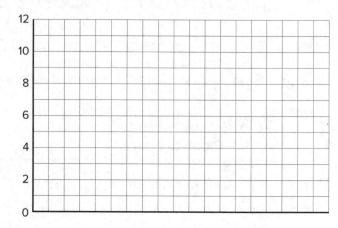

b. Think about how you would rearrange the hours so that each server gets a fair share. Then, on the second grid, draw a new graph to represent the rearranged hours. Be prepared to explain your reasoning.

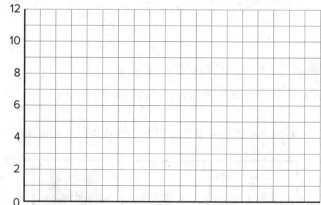

c. Based on your second drawing, what is the average or mean number of hours that the servers will work?

d. Explain why we can also find the mean by finding the value of 31 ÷ 5.

e. Which server will see the biggest change to work hours? Which server will see the least change?

Are you ready for more?

Server F, working 7 hours, offers to join the group of five servers, sharing their workload. If server F joins, will the mean number of hours worked increase or decrease? Explain how you know.

NAME _____ DATE _____ PERIOD _____

Activity
9.3 Getting to School

For the past 12 school days, Mai has recorded how long her bus rides to school take in minutes. The times she recorded are shown in the table.

Bus Ride Time (min)	9	8	6	9	10	7	6	12	9	8	10	8

1. Find the mean for Mai's data. Show your reasoning.

2. In this situation, what does the mean tell us about Mai's trip to school?

3. For 5 days, Tyler has recorded how long his walks to school take in minutes. The mean for his data is 11 minutes. Without calculating, predict if each of the data sets shown could be Tyler's. Explain your reasoning.

   • Data set A: 11, 8, 7, 9, 8

   • Data set B: 12, 7, 13, 9, 14

   • Data set C: 11, 20, 6, 9, 10

   • Data set D: 8, 10, 9, 11, 11

4. Determine which data set is Tyler's. Explain how you know.

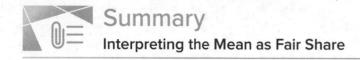
Sometimes a general description of a distribution does not give enough information, and a more precise way to talk about center or spread would be more useful. The **mean**, or **average**, is a number we can use to summarize a distribution.

We can think about the mean in terms of "fair share" or "leveling out." That is, a mean can be thought of as a number that each member of a group would have if all the data values were combined and distributed equally among the members.

For example, suppose there are 5 bottles which have the following amounts of water: 1 liter, 4 liters, 2 liters, 3 liters, and 0 liters.

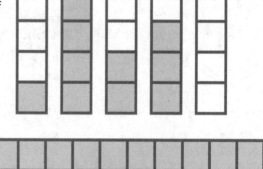

To find the mean, first we add up all of the values, which we can think of as putting all of the water together: $1 + 4 + 2 + 3 + 0 = 10$.

To find the "fair share," we divide the 10 liters equally into the 5 containers: $10 \div 5 = 2$.

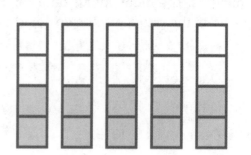

Suppose the quiz scores of a student are 70, 90, 86, and 94. We can find the mean (or average) score by finding the sum of the scores ($70 + 90 + 86 + 94 = 340$) and dividing the sum by four ($340 \div 4 = 85$). We can then say that the student scored, on average, 85 points on the quizzes.

In general, to find the mean of a data set with n values, we add all of the values and divide the sum by n.

Glossary

average

mean

NAME _____ DATE _____ PERIOD _____

Practice
Interpreting the Mean as Fair Share

1. A preschool teacher is rearranging four boxes of playing blocks so that each box contains an equal number of blocks. Currently Box 1 has 32 blocks, Box 2 has 18, Box 3 has 41, and Box 4 has 9.

   Select **all** the ways he could make each box have the same number of blocks.

   (A.) Remove all the blocks and make four equal piles of 25, then put each pile in one of the boxes.

   (B.) Remove 7 blocks from Box 1 and place them in Box 2.

   (C.) Remove 21 blocks from Box 3 and place them in Box 4.

   (D.) Remove 7 blocks from Box 1 and place them in Box 2, and remove 21 blocks from Box 3 and place them in Box 4.

   (E.) Remove 7 blocks from Box 1 and place them in Box 2, and remove 16 blocks from Box 3 and place them in Box 4.

2. In a round of mini-golf, Clare records the number of strokes it takes to hit the ball into the hole of each green.

   2   3   1   4   5   2   3   4   3

   She said that, if she redistributed the strokes on different greens, she could tell that her average number of strokes per hole is 3. Explain how Clare is correct.

3. Three sixth-grade classes raised $25.50, $49.75, and $37.25 for their classroom libraries. They agreed to share the money raised equally. What is each class's equal share? Explain or show your reasoning.

4. In her English class, Mai's teacher gives 4 quizzes each worth 5 points. After 3 quizzes, she has the scores 4, 3, and 4. What does she need to get on the last quiz to have a mean score of 4? Explain or show your reasoning.

5. An earthworm farmer examined two containers of a certain species of earthworms so that he could learn about their lengths. He measured 25 earthworms in each container and recorded their lengths in millimeters. Here are histograms of the lengths for each container. (Lesson 8-7)

    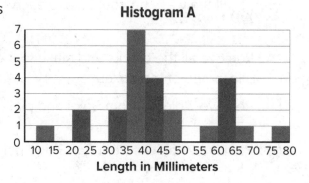

    a. Which container tends to have longer worms than the other container?

    b. For which container would 15 millimeters be a reasonable description of a typical length of the worms in the container?

    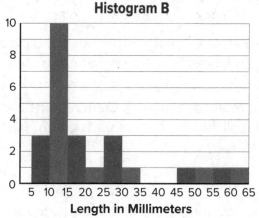

    c. If length is related to age, which container had the most young worms?

6. Diego thinks that $x = 3$ is a solution to the equation $x^2 = 16$. Do you agree? Explain or show your reasoning. (Lesson 6-15)

Lesson 8-10

Finding and Interpreting the Mean as the Balance Point

NAME _____ DATE _____ PERIOD _____

Learning Goal Let's look at another way to understand the mean of a data set.

Warm Up
10.1 Which One Doesn't Belong: Division

Which expression does not belong? Be prepared to explain your reasoning.

$$\frac{8+8+4+4}{4} \qquad \frac{10+10+4}{4} \qquad \frac{9+9+5+5}{4} \qquad \frac{6+6+6+6+6}{5}$$

Activity
10.2 Travel Times (Part 1)

Here is the data set from an earlier lesson showing how long it takes for Diego to walk to school, in minutes, over 5 days. The mean number of minutes is 11.

12   7   13   9   14

1. Represent Diego's data on a dot plot. Mark the location of the mean with a triangle.

2. The mean can also be seen as a **measure of center** that balances the points in a data set. If we find the distance between every point and the mean, add the distances on each side of the mean, and compare the two sums, we can see this balancing.

a. Record the distance between each point and 11 and its location relative to 11.

b. Sum of distances left of 11:_____
   Sum of distances right of 11:_____

   What do you notice about the two sums?

Time (minutes)	Distance from 11	Left of 11 or Right of 11?
12		
7		
13		
9		
14		

3. Can another point that is *not* the mean produce similar sums of distances?

   Let's investigate whether 10 can produce similar sums as those of 11.

a. Complete the table with the distance of each data point from 10.

b. Sum of distances left of 10:_____

   Sum of distances right of 10:_____

   What do you notice about the two sums?

Time (minutes)	Distance from 10	Left of 10 or Right of 10?
12		
7		
13		
9		
14		

4. Based on your work so far, explain why the mean can be considered a balance point for the data set.

NAME _____ DATE _____ PERIOD _____

Activity

10.3 Travel Times (Part 2)

1. Here are dot plots showing how long Diego's trips to school took in minutes—which you studied earlier—and how long Andre's trips to school took in minutes. The dot plots include the means for each data set, marked by triangles.

Diego's Travel Time in Minutes

Andre's Travel Time in Minutes

a. Which of the two data sets has a larger mean? In this context, what does a larger mean tell us?

b. Which of the two data sets has larger sums of distances to the left and right of the mean? What do these sums tell us about the variation in Diego's and Andre's travel times?

2. Here is a dot plot showing lengths of Lin's trips to school.

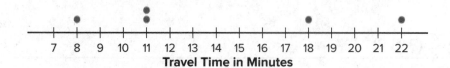

Travel Time in Minutes

a. Calculate the mean of Lin's travel times.

b. Complete the table with the distance between each point and the mean as well whether the point is to the left or right of the mean.

Time (minutes)	Distance from the Mean	Left or Right of the Mean?
22		
18		
11		
8		
11		

c. Find the sum of distances to the left of the mean and the sum of distances to the right of the mean.

d. Use your work to compare Lin's travel times to Andre's. What can you say about their average travel times? What about the variability in their travel times?

NAME _____ DATE _____ PERIOD _____

Summary

Finding and Interpreting the Mean as the Balance Point

The mean is often used as a **measure of center** of a distribution. This is because the mean of a distribution can be seen as the "balance point" for the distribution. Why is this a good way to think about the mean? Let's look at a very simple set of data on the number of cookies that each of eight friends baked.

19   20   20   21   21   22   22   23

Here is a dot plot showing the data set.

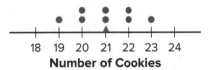

The distribution shown is completely symmetrical. The mean number of cookies is 21, because $(19 + 20 + 20 + 21 + 21 + 22 + 22 + 23) \div 8 = 21$. If we mark the location of the mean on the dot plot, we can see that the data points could balance at 21.

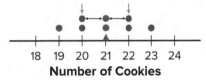

In this plot, each point on either side of the mean has a mirror image. For example, the two points at 20 and the two at 22 are the same distance from 21, but each pair is located on either side of 21. We can think of them as balancing each other around 21.

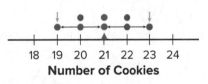

Similarly, the points at 19 and 23 are the same distance from 21 but are on either side of it. They, too, can be seen as balancing each other around 21.

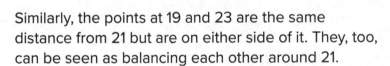

We can say that the distribution of the cookies has a center at 21 because that is its balance point, and that the eight friends, on average, baked 21 cookies.

Even when a distribution is not completely symmetrical, the distances of values below the mean, on the whole, balance the distances of values above the mean.

Glossary

measure of center

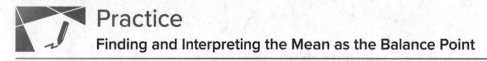

Practice
Finding and Interpreting the Mean as the Balance Point

1. On school days, Kiran walks to school. Here are the lengths of time, in minutes, for Kiran's walks on 5 school days.

    16   11   18   12   13

    a. Create a dot plot for Kiran's data.

    b. Without calculating, decide if 15 minutes would be a good estimate of the mean. If you think it is a good estimate, explain your reasoning. If not, give a better estimate and explain your reasoning.

    c. Calculate the mean for Kiran's data.

NAME _____ DATE _____ PERIOD _____

d. In the table, record the distance of each data point from the mean and its location relative to the mean.

Time (minutes)	Distance from the Mean	Left or Right of the Mean?
16		
11		
18		
12		
13		

e. Calculate the sum of all distances to the left of the mean, then calculate the sum of distances to the right of the mean. Explain how these sums show that the mean is a balance point for the values in the data set.

2. Noah scored 20 points in a game. Mai's score was 30 points. The mean score for Noah, Mai, and Clare was 40 points. What was Clare's score? Explain or show your reasoning.

3. Compare the numbers using >, <, or =. (Lesson 7-7)

   a. -2 _____ 3

   b. |-12| _____ |15|

   c. 3 _____ -4

   d. |15| _____ |-12|

   e. 7 _____ -11

   f. -4 _____ |5|

4. Respond to each of the following. (Lesson 7-3)

   a. Plot $\frac{2}{3}$ and $\frac{3}{4}$ on a number line.

   b. Is $\frac{2}{3} < \frac{3}{4}$, or is $\frac{3}{4} < \frac{2}{3}$? Explain how you know.

5. Select **all** the expressions that represent the total area of the large rectangle. (Lesson 6-10)

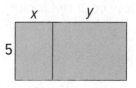

   (A.) $5(x + y)$

   (B.) $5 + xy$

   (C.) $5x + 5y$

   (D.) $2(5 + x + y)$

   (E.) $5xy$

Lesson 8-11

Deviation from the Mean

NAME _____ DATE _____ PERIOD _____

Learning Goal Let's study distances between data points and the mean and see what they tell us.

 ## Warm Up
11.1 Shooting Hoops (Part 1)

Elena, Jada, and Lin enjoy playing basketball during recess. Lately, they have been practicing free throws. They record the number of baskets they make out of 10 attempts. Here are their data sets for 12 school days.

Elena	4	5	1	6	9	7	2	8	3	3	5	7
Jada	2	4	5	4	6	6	4	7	3	4	8	7
Lin	3	6	6	4	5	5	3	5	4	6	6	7

1. Calculate the mean number of baskets each player made and compare the means. What do you notice?

2. What do the means tell us in this context?

Activity

11.2 Shooting Hoops (Part 2)

Here are the dot plots showing the number of baskets Elena, Jada, and Lin each made over 12 school days.

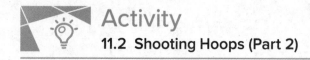

Number of Baskets Elena Made

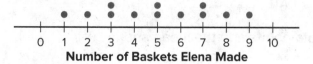

Number of Baskets Jada Made

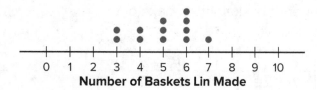

Number of Baskets Lin Made

1. On each dot plot, mark the location of the mean with a triangle (△). Then, contrast the dot plot distributions. Write 2–3 sentences to describe the shape and spread of each distribution.

2. Discuss the following questions with your group. Explain your reasoning.

   a. Would you say that all three students play equally well?

   b. Would you say that all three students play equally consistently?

   c. If you could choose one player to be on your basketball team based on their records, who would you choose?

NAME _____ DATE _____ PERIOD _____

Activity
11.3 Shooting Hoops (Part 3)

The tables show Elena, Jada, and Lin's basketball data from an earlier activity. Recall that the mean of Elena's data, as well as that of Jada and Lin's data, was 5.

1. Record the distance between each of Elena's scores and the mean.

Elena	4	5	1	6	9	7	2	8	3	3	5	7
Distance from 5	1			1								

Now find *the average of the distances* in the table. Show your reasoning and round your answer to the nearest tenth.

This value is the **mean absolute deviation (MAD)** of Elena's data.

Elena's MAD: _____

2. Find the mean absolute deviation of Jada's data. Round it to the nearest tenth.

Jada	2	4	5	4	6	6	4	7	3	4	8	7
Distance from 5												

Jada's MAD: _____

3. Find the mean absolute deviation of Lin's data. Round it to the nearest tenth.

Lin	3	6	6	4	5	5	3	5	4	6	6	7
Distance from 5												

Lin's MAD: _____

4. Compare the MADs and dot plots of the three students' data. Do you see a relationship between each student's MAD and the distribution on her dot plot? Explain your reasoning.

Number of Baskets Elena Made

Number of Baskets Jada Made

Number of Baskets Lin Made

Are you ready for more?

Invent another data set that also has a mean of 5 but has an MAD greater than 2. Remember, the values in the data set must be whole numbers from 0 to 10.

Activity

11.4 Game of 22

Your teacher will give your group a deck of cards. Shuffle the cards and put the deck face down on the playing surface.

- To play: Draw 3 cards and add up the values. An ace is a 1. A jack, queen, and king are each worth 10. Cards 2–10 are each worth their face value. If your sum is anything other than 22 (either above or below 22), say: "My sum deviated from 22 by _____ ," or "My sum was off from 22 by _____ ."

- To keep score: Record each sum and each distance from 22 in the table. After five rounds, calculate the average of the distances. The player with the lowest average distance from 22 wins the game.

Player A	Round 1	Round 2	Round 3	Round 4	Round 5
Sum of Cards					
Distance from 22					

Average distance from 22: _____

Player B	Round 1	Round 2	Round 3	Round 4	Round 5
Sum of Cards					
Distance from 22					

Average distance from 22: _____

Player C	Round 1	Round 2	Round 3	Round 4	Round 5
Sum of Cards					
Distance from 22					

Average distance from 22: _____

Whose average distance from 22 is the smallest? Who won the game?

Summary

Deviation from the Mean

We use the mean of a data set as a measure of center of its distribution, but two data sets with the same mean could have very different distributions.

This dot plot shows the weights, in grams, of 22 cookies. The mean weight is 21 grams. All the weights are within 3 grams of the mean, and most of them are even closer. These cookies are all fairly close in weight.

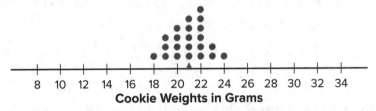

This dot plot shows the weights, in grams, of a different set of 30 cookies.

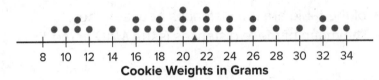

The mean weight for this set of cookies is also 21 grams, but some cookies are half that weight and others are one-and-a-half times that weight. There is a lot more variability in the weight.

There is a number we can use to describe how far away, or how spread out, data points generally are from the mean. This *measure of spread* is called the **mean absolute deviation (MAD)**.

Here the MAD tells us how far cookie weights typically are from 21 grams. To find the MAD, we find the distance between each data value and the mean, and then calculate the mean of those distances.

For instance, the point that represents 18 grams is 3 units away from the mean of 21 grams. We can find the distance between each point and the mean of 21 grams and organize the distances into a table, as shown.

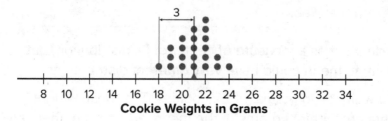

Cookie Weights in Grams

Weight (grams)	18	19	19	19	20	20	20	20	21	21	21	21	21	22	22	22	22	22	22	23	23	24
Distance from Mean	3	2	2	2	1	1	1	1	0	0	0	0	0	1	1	1	1	1	1	2	2	3

The values in the first row of the table are the cookie weights for the first set of cookies. Their mean, 21 grams, is the *mean of the cookie weights*.

The values in the second row of the table are the *distances* between the values in the first row and 21. The mean of these distances is the *MAD of the cookie weights*.

What can we learn from the averages of these distances once they are calculated?

- In the first set of cookies, the distances are all between 0 and 3. The MAD is 1.2 grams, which tells us that the cookie weights are typically within 1.2 grams of 21 grams. We could say that a typical cookie weighs between 19.8 and 22.2 grams.

- In the second set of cookies, the distances are all between 0 and 13. The MAD is 5.6 grams, which tells us that the cookie weights are typically within 5.6 grams of 21 grams. We could say a typical cookie weighs between 15.4 and 26.6 grams.

The MAD is also called a *measure of the variability* of the distribution. In these examples, it is easy to see that a higher MAD suggests a distribution that is more spread out, showing more variability.

Glossary
_____

mean absolute deviation (MAD)

NAME _____ DATE _____ PERIOD _____

Practice
Deviation from the Mean

1. Han recorded the number of pages that he read each day for five days. The dot plot shows his data.

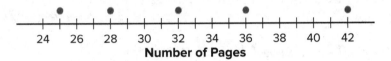

Number of Pages

a. Is 30 pages a good estimate of the mean number of pages that Han read each day? Explain your reasoning.

b. Find the mean number of pages that Han read during the five days. Draw a triangle to mark the mean on the dot plot.

c. Use the dot plot and the mean to complete the table.

Number of Pages	Distance from Mean	Left or Right of Mean
25		
28		
32		
36		
42		

d. Calculate the mean absolute deviation (MAD) of the data. Explain or show your reasoning.

2. Ten sixth-grade students recorded the amounts of time each took to travel to school. The dot plot shows their travel times.

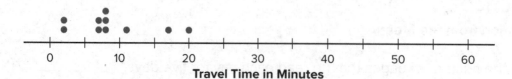

Travel Time in Minutes

The mean travel time for these students is approximately 9 minutes. The MAD is approximately 4.2 minutes.

a. Which number of minutes—9 or 4.2—is a typical amount of time for the ten sixth-grade students to travel to school? Explain your reasoning.

b. Based on the mean and MAD, Jada believes that travel times between 5 and 13 minutes are common for this group. Do you agree? Explain your reasoning.

c. A different group of ten sixth-grade students also recorded their travel times to school. Their mean travel time was also 9 minutes, but the MAD was about 7 minutes. What could the dot plot of this second data set be? Describe or draw how it might look.

3. In an archery competition, scores for each round are calculated by averaging the distance of 3 arrows from the center of the target.

An archer has a mean distance of 1.6 inches and a MAD distance of 1.3 inches in the first round. In the second round, the archer's arrows are farther from the center but are more consistent. What values for the mean and MAD would fit this description for the second round? Explain your reasoning.

Lesson 8-12

Using Mean and MAD to Make Comparisons

NAME _____ DATE _____ PERIOD _____

Learning Goal Let's use mean and MAD to describe and compare distributions.

Warm Up
12.1 Number Talk: Decimal Division

Find the value of each expression mentally.

1. 42 ÷ 12   **2.** 2.4 ÷ 12   **3.** 44.4 ÷ 12   **4.** 46.8 ÷ 12

Activity
12.2 Which Player Would You Choose?

1. Andre and Noah joined Elena, Jada, and Lin in recording their basketball scores. They all recorded their scores in the same way: the number of baskets made out of 10 attempts. Each collected 12 data points.

Andre's mean number of baskets was 5.25, and his MAD was 2.6. Noah's mean number of baskets was also 5.25, but his MAD was 1.

Here are two dot plots that represent the two data sets. The triangle indicates the location of the mean.

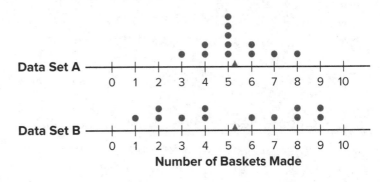

a. Without calculating, decide which dot plot represents Andre's data and which represents Noah's. Explain how you know.

b. If you were the captain of a basketball team and could use one more player on your team, would you choose Andre or Noah? Explain your reasoning.

2. An eighth-grade student decided to join Andre and Noah and kept track of his scores. His data set is shown here. The mean number of baskets he made is 6.

Eighth-Grade Student	6	5	4	7	6	5	7	8	5	6	5	8
Distance from 6												

a. Calculate the MAD. Show your reasoning.

b. Draw a dot plot to represent his data and mark the location of the mean with a triangle (△).

c. Compare the eighth-grade student's mean and MAD to Noah's mean and MAD. What do you notice?

d. Compare their dot plots. What do you notice about the distributions?

e. What can you say about the two players' shooting accuracy and consistency?

NAME _____ DATE _____ PERIOD _____

Are you ready for more?

Invent a data set with a mean of 7 and an MAD of 1.

Activity

12.3 Swimmers Over the Years

In 1984, the mean age of swimmers on the U.S. women's swimming team was 18.2 years and the MAD was 2.2 years. In 2016, the mean age of the swimmers was 22.8 years, and the MAD was 3 years.

1. How has the average age of the women on the U.S. swimming team changed from 1984 to 2016? Explain your reasoning.

2. Are the swimmers on the 1984 team closer in age to one another than the swimmers on the 2016 team are to one another? Explain your reasoning.

3. Here are dot plots showing the ages of the women on the U.S. swimming team in 1984 and in 2016. Use them to make two other comments about how the women's swimming team has changed over the years.

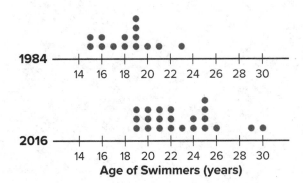

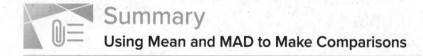

Summary

Using Mean and MAD to Make Comparisons

Sometimes two distributions have different means but the same MAD.

Pugs and beagles are two different dog breeds. The dot plot shows two sets of weight data—one for pugs and the other for beagles.

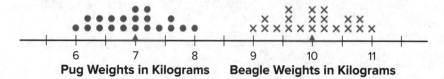

- The mean weight for pugs is 7 kilograms, and the MAD is 0.5 kilogram.

- The mean weight for beagles is 10 kilograms, and the MAD is 0.5 kilogram.

We can say that, in general, the beagles are heavier than the pugs. A typical weight for the beagles in this group is about 3 kilograms heavier than a typical weight for the pugs.

The variability of pug weights, however, is about the same as the variability of beagle weights. In other words, the weights of pugs and the weights of beagles are equally spread out.

NAME _____ DATE _____ PERIOD _____

Practice

Using Mean and MAD to Make Comparisons

1. The dot plots show the amounts of time that ten U.S. students and ten Australian students took to get to school. Which statement is true about the MAD of the Australian data set?

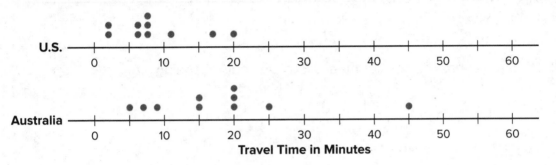

Travel Time in Minutes

(A.) It is significantly less than the MAD of the U.S. data set.

(B.) It is exactly equal to the MAD of the U.S. data set.

(C.) It is approximately equal to the MAD of the U.S. data set.

(D.) It is significantly greater than the MAD of the U.S. data set.

2. The dot plots show the amounts of time that ten South African students and ten Australian students took to get to school. Without calculating, answer the questions.

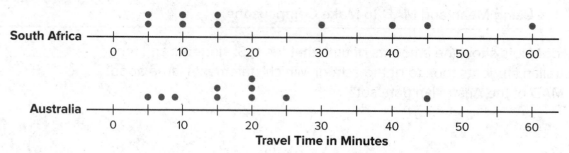

a. Which data set has the smaller mean? Explain your reasoning.

b. Which data set has the smaller MAD? Explain your reasoning.

c. What does a smaller mean tell us in this context?

d. What does a smaller MAD tell us in this context?

NAME _____ DATE _____ PERIOD _____

3. Two high school basketball teams have identical records of 15 wins and 2 losses. Sunnyside High School's mean score is 50 points and its MAD is 4 points. Shadyside High School's mean score is 60 points and its MAD is 15 points. Lin read the records of each team's score. She likes the team that had nearly the same score for every game it played. Which team do you think Lin likes? Explain your reasoning.

4. Jada thinks the perimeter of this rectangle can be represented with the expression $a + a + b + b$. Andre thinks it can be represented with $2a + 2b$. Do you agree with either of them? Explain your reasoning. **(Lesson 6-8)**

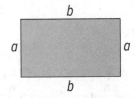

5. Draw a number line. (Lesson 7-3)

    a. Plot and label three numbers between -2 and -8 (not including -2 and -8). Answers vary.

    b. Use the numbers you plotted and the symbols < and > to write three inequality statements. Answers vary.

6. Adult elephant seals generally weigh about 5,500 pounds. If you weighed 5 elephant seals, would you expect each seal to weigh exactly 5,500 pounds? Explain your reasoning. (Lesson 8-2)

Lesson 8-13

The Median of a Data Set

NAME _____ DATE _____ PERIOD _____

Learning Goal Let's explore the median of a data set and what it tells us.

Warm Up
13.1 The Plot of the Story

1. Here are two dot plots and two stories. Match each story with a dot plot that could represent it. Be prepared to explain your reasoning.

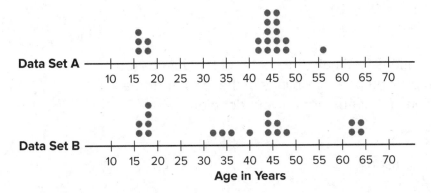

- Twenty people—high school students, teachers, and invited guests—attended a rehearsal for a high school musical. The mean age was 38.5 years and the MAD was 16.5 years.

- High school soccer team practice is usually watched by supporters of the players. One evening, twenty people watched the team practice. The mean age was 38.5 years and the MAD was 12.7 years.

2. Another evening, twenty people watched the soccer team practice. The mean age was similar to that from the first evening, but the MAD was greater (about 20 years). Make a dot plot that could illustrate the distribution of ages in this story.

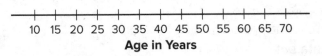

Activity

13.2 Siblings in the House

Here is data that shows the numbers of siblings of ten students in Tyler's class.

1   0   2   1   7   0   2   0   1   10

1. Represent the data shown with a dot plot.

2. Without making any calculations, estimate the center of the data based on your dot plot. What is a typical number of siblings of these sixth-grade students? Mark the location of that number on your dot plot.

3. Find the mean. Show your reasoning.

4. **a.** How does the mean compare to the value that you marked on the dot plot as a typical number of siblings? (Is it a little larger, a lot larger, exactly the same, a little smaller, or a lot smaller than your estimate?)

   b. Do you think the mean summarizes the data set well? Explain your reasoning.

Are you ready for more?

Invent a data set with a mean that is significantly lower than what you would consider a typical value for the data set.

NAME _____ DATE _____ PERIOD _____

 Activity
13.3 Finding the Middle

1. Your teacher will give you an index card. Write your first and last names on the card. Then record the total number of letters in your name. After that, pause for additional instructions from your teacher.

2. Here is the data set on numbers of siblings from an earlier activity.

   1   0   2   1   7   0   2   0   1   10

   a. Sort the data from least to greatest, and then find the **median**.

   b. In this situation, do you think the median is a good measure of a typical number of siblings for this group? Explain your reasoning.

3. Here is the dot plot showing the travel time, in minutes, of Elena's bus rides to school.

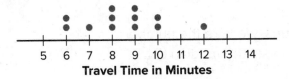
Travel Time in Minutes

   a. Find the median travel time. Be prepared to explain your reasoning.

   b. What does the median tell us in this context?

The **median** is another measure of center of a distribution. It is the middle value in a data set when values are listed in order. Half of the values in a data set are less than or equal to the median, and half of the values are greater than or equal to the median.

To find the median, we order the data values from least to greatest and find the number in the middle.

Suppose we have 5 dogs whose weights, in pounds, are shown in the table. The median weight for this group of dogs is 32 pounds because three dogs weigh less than or equal to 32 pounds and three dogs weigh greater than or equal to 32 pounds.

20	25	32	40	55

Now suppose we have 6 cats whose weights, in pounds, are as shown in the table. Notice that there are *two* values in the middle: 7 and 8.

4	6	7	8	10	10

The median weight must be between 7 and 8 pounds, because half of the cats weigh less or equal to 7 pounds and half of the cats weigh greater than or equal to 8 pounds.

In general, when we have an even number of values, we take the number exactly in between the two middle values. In this case, the median cat weight is 7.5 pounds because $(7 + 8) \div 2 = 7.5$.

Glossary

median

NAME _____ DATE _____ PERIOD _____

Practice
The Median of a Data Set

1. Here is data that shows a student's scores for 10 rounds of a video game.

130 150 120 170 130 120 160 160 190 140

What is the median score?

(A.) 125

(C.) 147

(B.) 145

(D.) 150

2. When he sorts the class's scores on the last test, the teacher notices that exactly 12 students scored better than Clare and exactly 12 students scored worse than Clare. Does this mean that Clare's score on the test is the median? Explain your reasoning.

3. The medians of the following dot plots are 6, 12, 13, and 15, but not in that order. Match each dot plot with its median.

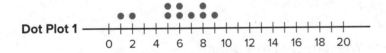

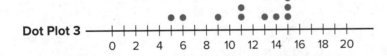

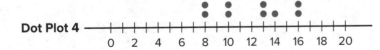

4. Invent a data set with five numbers that has a mean of 10 and a median of 12.

5. Ten sixth-grade students reported the hours of sleep they get on nights before a school day. Their responses are recorded in the dot plot.

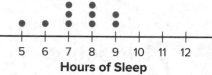

Looking at the dot plot, Lin estimated the mean number of hours of sleep to be 8.5 hours. Noah's estimate was 7.5 hours. Diego's estimate was 6.5 hours. Which estimate do you think is best? Explain how you know. (Lesson 8-10)

6. In one study of wild bears, researchers measured the weights, in pounds, of 143 wild bears that ranged in age from newborn to 15 years old. The data were used to make this histogram. (Lesson 8-8)

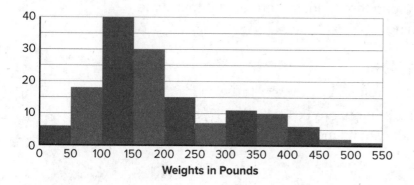

Weights in Pounds

a. What can you say about the heaviest bear in this group?

b. What is a typical weight for the bears in this group?

c. Do more than half of the bears in this group weigh less than 250 pounds?

d. If weight is related to age, with older bears tending to have greater body weights, would you say that there were more old bears or more young bears in the group? Explain your reasoning.

Lesson 8-14

Comparing Mean and Median

NAME _____ DATE _____ PERIOD _____

Learning Goal Let's compare the mean and median of data sets.

Warm Up
14.1 Heights of Presidents

Here are two dot plots. The first dot plot shows the heights of the first 22 U.S. presidents. The second dot plot shows the heights of the next 22 presidents.

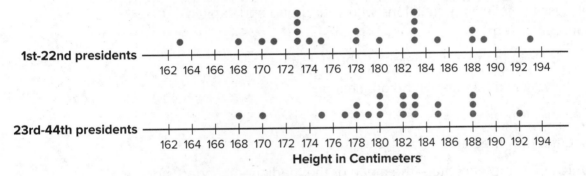

Based on the two dot plots, decide if you agree or disagree with each of the following statements. Be prepared to explain your reasoning.

1. The median height of the first 22 presidents is 178 centimeters.

2. The mean height of the first 22 presidents is about 183 centimeters.

3. A typical height for a president in the second group is about 182 centimeters.

4. U.S. presidents have become taller over time.

5. The heights of the first 22 presidents are more alike than the heights of the second 22 presidents.

6. The MAD of the second data set is greater than the MAD of the first set.

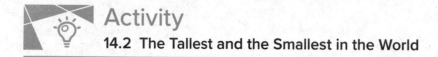

Activity

14.2 The Tallest and the Smallest in the World

Your teacher will provide the height data for your class. Use the data to complete the following questions.

1. Find the mean height of your class in centimeters.

2. Find the median height in centimeters. Show your reasoning.

3. Suppose that the world's tallest adult, who is 251 centimeters tall, joined your class.

a. Discuss the following questions with your group and explain your reasoning.

- How would the mean height of the class change?

- How would the median height change?

b. Find the new mean.

c. Find the new median.

d. Which measure of center—the mean or the median—changed more when this new person joined the class? Explain why the value of one measure changed more than the other.

4. The world's smallest adult is 63 centimeters tall. Suppose that the world's tallest and smallest adults both joined your class.

a. Discuss the following questions with your group and explain your reasoning.

- How would the mean height of the class change from the original mean?

- How would the median height change from the original median?

b. Find the new mean.

c. Find the new median.

d. How did the measures of center—the mean and the median—change when these two people joined the class? Explain why the values of the mean and median changed the way they did.

NAME _____ DATE _____ PERIOD _____

 Activity
14.3 Mean or Median?

1. Your teacher will give you six cards. Each has either a dot plot or a histogram. Sort the cards into *two* piles based on the distributions shown. Be prepared to explain your reasoning.

2. Discuss your sorting decisions with another group. Did you have the same cards in each pile? If so, did you use the same sorting categories? If not, how are your categories different?

   Pause here for a class discussion.

3. Use the information on the cards to answer the following questions.

   Card A: What is a typical age of the dogs being treated at the animal clinic?

   Card B: What is a typical number of people in the Irish households?

   Card C: What is a typical travel time for the New Zealand students?

   Card D: Would 15 years old be a good description of a typical age of the people who attended the birthday party?

   Card E: Is 15 minutes or 24 minutes a better description of a typical time it takes the students in South Africa to get to school?

   Card F: Would 21.3 years old be a good description of a typical age of the people who went on a field trip to Washington, D.C.?

4. How did you decide which measure of center to use for the dot plots on Cards A–C? What about for those on Cards D–F?

Are you ready for more?

Most teachers use the mean to calculate a student's final grade, based on that student's scores on tests, quizzes, homework, projects, and other graded assignments.

Diego thinks that the median might be a better way to measure how well a student did in a course. Do you agree with Diego? Explain your reasoning.

Both the mean and the median are ways of measuring the center of a distribution. They tell us slightly different things, however.

The dot plot shows the weights of 30 cookies. The mean weight is 21 grams (marked with a triangle). The median weight is 20.5 grams (marked with a diamond).

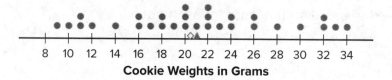

Cookie Weights in Grams

The mean tells us that if the weights of all cookies were distributed so that each one weighed the same, that weight would be 21 grams. We could also think of 21 grams as a balance point for the weights of all of the cookies in the set.

The median tells us that half of the cookies weigh more than 20.5 grams and half weigh less than 20.5 grams. In this case, both the mean and the median could describe a typical cookie weight because they are fairly close to each other and to most of the data points.

Here is a different set of 30 cookies. It has the same mean weight as the first set, but the median weight is 23 grams.

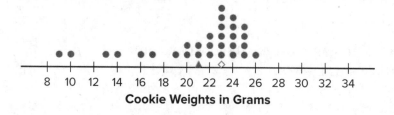

Cookie Weights in Grams

In this case, the median is closer to where most of the data points are clustered and is therefore a better measure of center for this distribution. That is, it is a better description of a typical cookie weight. The mean weight is influenced (in this case, pulled down) by a handful of much smaller cookies, so it is farther away from most data points.

In general, when a distribution is symmetrical or approximately symmetrical, the mean and median values are close. But when a distribution is not roughly symmetrical, the two values tend to be farther apart.

NAME _____ DATE _____ PERIOD _____

Practice
Comparing Mean and Median

1. Here is a dot plot that shows the ages of teachers at a school.

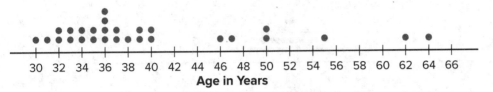

   Which of these statements is true of the data set shown in the dot plot?

   (A.) The mean is less than the median.

   (B.) The mean is approximately equal to the median.

   (C.) The mean is greater than the median.

   (D.) The mean cannot be determined.

2. Priya asked each of five friends to attempt to throw a ball in a trash can until they succeeded. She recorded the number of unsuccessful attempts made by each friend as: 1, 8, 6, 2, 4. Priya made a mistake: The 8 in the data set should have been 18.

   How would changing the 8 to 18 affect the mean and median of the data set?

   (A.) The mean would decrease and the median would not change.

   (B.) The mean would increase and the median would not change.

   (C.) The mean would decrease and the median would increase.

   (D.) The mean would increase and the median would increase.

3. In his history class, Han's homework scores are:

   100   100   100   100   95   100   90   100   0

   The history teacher uses the mean to calculate the grade for homework. Write an argument for Han to explain why median would be a better measure to use for his homework grades.

4. The dot plots show how much time, in minutes, students in a class took to complete each of five different tasks. Select **all** the dot plots of tasks for which the mean time is approximately equal to the median time.

(A.)
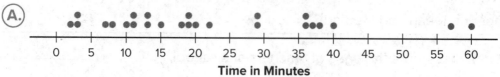

(B.)
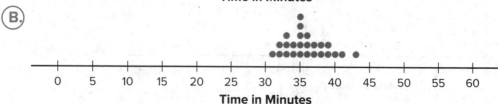

(C.)
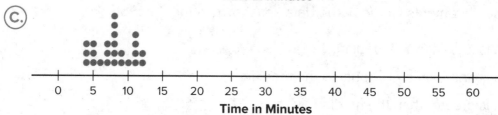

(D.)
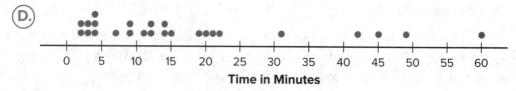

(E.)
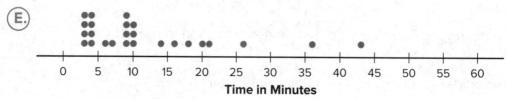

5. Zookeepers recorded the ages, weights, genders, and heights of the 10 pandas at their zoo. Write two statistical questions that could be answered using these data sets. (Lesson 8-2)

6. Here is a set of coordinates. Draw and label an appropriate pair of axes and plot the points. $A = (1, 0)$, $B = (0, 0.5)$, $C = (4, 3.5)$, $D = (1.5, 0.5)$ (Lesson 7-12)

Lesson 8-15

Quartiles and Interquartile Range

NAME _____ DATE _____ PERIOD _____

Learning Goal Let's look at other measures for describing distributions.

Warm Up
15.1 Notice and Wonder: Two Parties

Here are dot plots that show the ages of people at two different parties. The mean of each distribution is marked with a triangle.

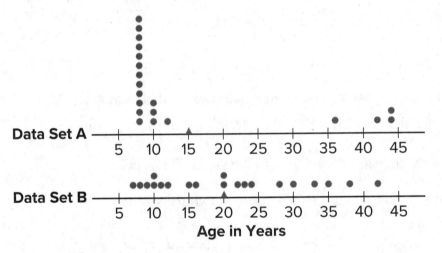

What do you notice and what do you wonder about the distributions in the two dot plots?

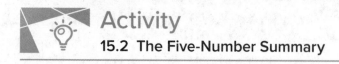

Activity

15.2 The Five-Number Summary

Here are the ages of the people at one party, listed from least to greatest.

7 8 9 10 10 11 12 15 16 20 20 22 23 24 28 30 33 35 38 42

1. a. Find the median of the data set and label it "50th percentile." This splits the data into an upper half and a lower half.

   b. Find the middle value of the *lower* half of the data, without including the median. Label this value "25th percentile."

   c. Find the middle value of the *upper* half of the data, without including the median. Label this value "75th percentile."

2. You have split the data set into four pieces. Each of the three values that split the data is called a **quartile**.

   • We call the 25th percentile the *first quartile*. Write "Q1" next to that number.

   • The median can be called the *second quartile*. Write "Q2" next to that number.

   • We call the 75th percentile the *third quartile*. Write "Q3" next to that number.

3. Label the lowest value in the set "minimum" and the greatest value "maximum."

4. The values you have identified make up the *five-number summary* for the data set. Record them here.

   minimum: _____ Q1: _____ Q2: _____ Q3: _____ maximum: _____

5. The median of this data set is 20. This tells us that half of the people at the party were 20 years old or younger, and the other half were 20 or older. What do each of these other values tell us about the ages of the people at the party?

   a. the third quartile

   b. the minimum

   c. the maximum

NAME _____ DATE _____ PERIOD _____

Are you ready for more?

There was another party where 21 people attended. Here is the five-number summary of their ages.

minimum: __5__   Q1: __6__   Q2: __27__   Q3: __32__   maximum: __60__

1. Do you think this party had more children or fewer children than the earlier one? Explain your reasoning.

2. Were there more children or adults at this party? Explain your reasoning.

Activity

15.3 Range and Interquartile Range

1. Here is a dot plot that shows the lengths of Elena's bus rides to school, over 12 days.

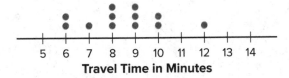

Travel Time in Minutes

Write the five-number summary for this data set. Show your reasoning.

2. The **range** is one way to describe the *spread* of values in a data set. It is the difference between the maximum and minimum. What is the range of Elena's travel times?

3. Another way to describe the spread of values in a data set is the **interquartile range (IQR)**. It is the difference between the upper quartile and the lower quartile.

   a. What is the interquartile range (IQR) of Elena's travel times?

   b. What fraction of the data values are between the lower and upper quartiles?

4. Here are two more dot plots.

Data Set A

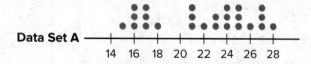

Data Set B

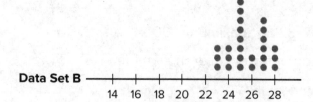

Without doing any calculations, predict:

a. Which data set has the smaller range?

b. Which data set has the smaller IQR?

5. Check your predictions by calculating the range and IQR for the data in each dot plot.

NAME _____ DATE _____ PERIOD _____

Summary
Quartiles and Interquartile Range

Earlier we learned that the mean is a measure of the center of a distribution and the MAD is a measure of the variability (or spread) that goes with the mean. There is also a measure of spread that goes with the median. It is called the interquartile range (IQR).

Finding the IQR involves splitting a data set into fourths. Each of the three values that splits the data into fourths is called a **quartile**.

- The median, or second quartile (Q2), splits the data into two halves.

- The first quartile (Q1) is the middle value of the lower half of the data.

- The third quartile (Q3) is the middle value of the upper half of the data.

For example, here is a data set with 11 values.

12 19 20 21 22 33 34 35 40 40 49
       Q1   **Q2**   **Q3**

- The median is 33.

- The first quartile is 20. It is the median of the numbers less than 33.

- The third quartile is 40. It is the median of the numbers greater than 33.

The difference between the maximum and minimum values of a data set is the **range**. The difference between Q3 and Q1 is the **interquartile range (IQR)**. Because the distance between Q1 and Q3 includes the middle two-fourths of the distribution, the values between those two quartiles are sometimes called the *middle half of the data*.

The bigger the IQR, the more spread out the middle half of the data values are. The smaller the IQR, the closer together the middle half of the data values are. This is why we can use the IQR as a measure of spread.

A *five-number summary* can be used to summarize a distribution. It includes the minimum, first quartile, median, third quartile, and maximum of the data set. For the previous example, the five-number summary is 12, 20, 33, 40, and 49. These numbers are marked with diamonds on the dot plot.

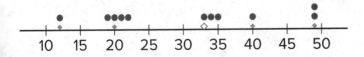

Different data sets can have the same five-number summary. For instance, here is another data set with the same minimum, maximum, and quartiles as the previous example.

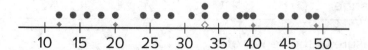

Glossary

interquartile range (IQR)
quartile
range

NAME _____ DATE _____ PERIOD _____

Practice
Quartiles and Interquartile Range

1. Suppose that there are 20 numbers in a data set and that they are all different.

   a. How many of the values in this data set are between the first quartile and the third quartile?

   b. How many of the values in this data set are between the first quartile and the median?

2. In a word game, 1 letter is worth 1 point. This dot plot shows the scores for 20 common words.

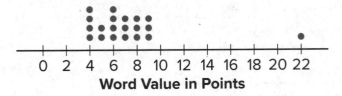

   Word Value in Points

   a. What is the median score?

   b. What is the first quartile (Q1)?

   c. What is the third quartile (Q3)?

   d. What is the interquartile range (IQR)?

3. Mai and Priya each played 10 games of bowling and recorded the scores. Mai's median score was 120, and her IQR was 5. Priya's median score was 118, and her IQR was 15. Whose scores probably had less variability? Explain how you know.

4. Here are five dot plots that show the amounts of time that ten sixth-grade students in five countries took to get to school. Match each dot plot with the appropriate median and IQR.

Dot Plots

Median & IQR

a.

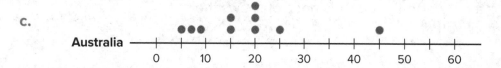
United States

1. Median: 175, IQR: 11

2. Median: 15, IQR: 30

3. Median: 8, IQR: 4

b.

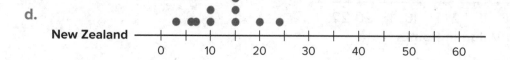
Canada

4. Median: 7, IQR: 10

5. Median: 12.5, IQR: 8

c.

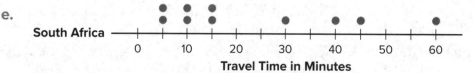
Australia

d.

New Zealand

e.

South Africa

Travel Time in Minutes

5. Draw and label an appropriate pair of axes and plot the points. A = (10, 50), B = (30, 25), C = (0, 30), D = (20, 35) (Lesson 7-12)

6. There are 20 pennies in a jar. If 16% of the coins in the jar are pennies, how many coins are there in the jar? (Lesson 6-7)

Lesson 8-16

Box Plots

NAME _____ DATE _____ PERIOD _____

Learning Goal Let's explore how box plots can help us summarize distributions.

 ## Warm Up
16.1 Notice and Wonder: Puppy Weights

Here are the birth weights, in ounces, of all the puppies born at a kennel in the past month.

13	14	15	15	16	16	16	16	17	17	17	17	17
17	17	18	18	18	18	18	18	18	18	19	20	

What do you notice and wonder about the distribution of the puppy weights?

 ## Activity
16.2 Human Box Plot

Your teacher will give you the data on the lengths of names of students in your class. Write the five-number summary by finding the data set's minimum, Q1, Q2, Q3, and the maximum.

Pause for additional instructions from your teacher.

Twenty people participated in a study about blinking. The number of times each person blinked while watching a video for one minute was recorded. The data values are shown here, in order from smallest to largest.

3   6   8   11   11   13   14   14   14   14
16   18   20   20   20   22   24   32   36   51

1. a. Use the grid and axis to make a dot plot of this data set.

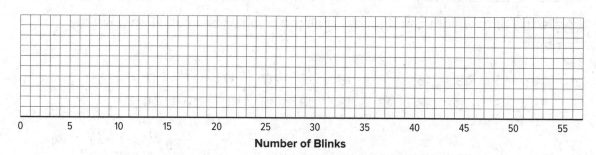

Number of Blinks

   b. Find the median (Q2) and mark its location on the dot plot.

   c. Find the first quartile (Q1) and the third quartile (Q3). Mark their locations on the dot plot.

   d. What are the minimum and maximum values?

2. A **box plot** can be used to represent the five-number summary graphically. Let's draw a box plot for the number-of-blinks data. On the grid, *above* the dot plot:

   a. Draw a box that extends from the first quartile (Q1) to the third quartile (Q3). Label the quartiles.

   b. At the median (Q2), draw a vertical line from the top of the box to the bottom of the box. Label the median.

   c. From the left side of the box (Q1), draw a horizontal line (a whisker) that extends to the minimum of the data set. On the right side of the box (Q3), draw a similar line that extends to the maximum of the data set.

NAME _____ DATE _____ PERIOD _____

3. You have now created a box plot to represent the number of blinks data. What fraction of the data values are represented by each of these elements of the box plot?

   a. The left whisker

   b. The box

   c. The right whisker

Are you ready for more?

Suppose there were some errors in the data set: the smallest value should have been 6 instead of 3, and the largest value should have been 41 instead of 51. Determine if any part of the five-number summary would change. If you think so, describe how it would change. If not, explain how you know.

Summary
Box Plots

A **box plot** represents the five-number summary of a data set.

It shows:

- the first quartile (Q1) and the third quartile (Q3) as the left and right sides of a rectangle or a box.

- the median (Q2) is shown as a vertical segment inside the box.

- on the left side, a horizontal line segment—a "whisker"—extends from Q1 to the minimum value.

- on the right, a whisker extends from Q3 to the maximum value.

- the rectangle in the middle represents the middle half of the data. Its width is the IQR.

- the whiskers represent the bottom quarter and top quarter of the data set.

Earlier we saw dot plots representing the weights of pugs and beagles. The box plots for these data sets are shown above the corresponding dot plots.

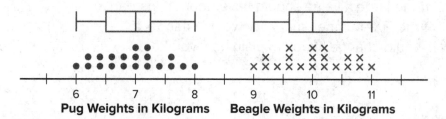

We can tell from the box plots that, in general, the pugs in the group are lighter than the beagles: the median weight of pugs is 7 kilograms and the median weight of beagles is 10 kilograms. Because the two box plots are on the same scale and the rectangles have similar widths, we can also tell that the IQRs for the two breeds are very similar. This suggests that the variability in the beagle weights is very similar to the variability in the pug weights.

Glossary

box plot

NAME _____ DATE _____ PERIOD _____

Practice
Box Plots

1. Each student in a class recorded how many books they read during the summer. Here is a box plot that summarizes their data.

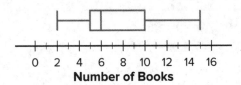

Number of Books

   a. What is the greatest number of books read by a student in this group?

   b. What is the median number of books read by the students?

   c. What is the interquartile range (IQR)?

2. Use this five-number summary to draw a box plot. All values are in seconds.

   - Minimum: 40
   - First quartile (Q1): 45
   - Median: 48

   - Third quartile (Q3): 50
   - Maximum: 60

3. The data show the number of hours per week that each of 13 seventh-grade students spent doing homework. Create a box plot to summarize the data.

   3   10   12   4   7   9   5   5   11   11   5   12   11

4. The box plot displays the data on the response times of 100 mice to seeing a flash of light. How many mice are represented by the rectangle between 0.5 and 1 second?

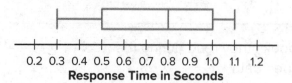

0.2 0.3 0.4 0.5 0.6 0.7 0.8 0.9 1.0 1.1 1.2
Response Time in Seconds

5. Here is a dot plot that represents a data set. Explain why the mean of the data set is greater than its median. **(Lesson 8-14)**

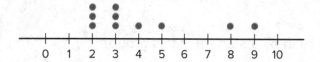

0 1 2 3 4 5 6 7 8 9 10

6. Jada earns money from babysitting, walking her neighbor's dogs, and running errands for her aunt. Every four weeks, she combines her earnings and divides them into three equal parts—one for spending, one for saving, and one for donating to a charity. Jada donated $26.00 of her earnings from the past four weeks to charity.

How much could she have earned from each job? Make two lists of how much she could have earned from the three jobs during the past four weeks. **(Lesson 8-9)**

Lesson 8-17

Using Box Plots

NAME _____ DATE _____ PERIOD _____

Learning Goal Let's use box plots to make comparisons.

Warm Up
17.1 Hours of Slumber

Ten sixth-grade students were asked how much sleep, in hours, they usually get on a school night. Here is the five-number summary of their responses.

- Minimum: 5 hours
- First quartile: 7 hours
- Median: 7.5 hours
- Third quartile: 8 hours
- Maximum: 9 hours

1. On the grid, draw a box plot for this five-number summary.

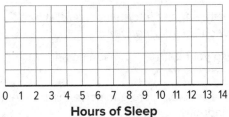

0 1 2 3 4 5 6 7 8 9 10 11 12 13 14

Hours of Sleep

2. What questions could be answered by looking at this box plot?

Activity

17.2 Info Gap Sea Turtles

Your teacher will give you either a
Problem Card or a Data Card about
sea turtles that nest on the
Outer Banks of North Carolina.
Do not show or read your card to
your partner.

If your teacher gives you the *problem card*:	If your teacher gives you the *data card*:

1. Silently read your card and think about what information you need to be able to answer the question.
 	
2. Ask your partner for the specific information that you need.
 	
3. Explain how you are using the information to solve the problem.
 	
   Continue to ask questions until you have enough information to solve the problem.
 	
4. Share the *problem card* and solve the problem independently.
 	
5. Read the *data card* and discuss your reasoning.
	1. Silently read your card.
	
	2. Ask your partner *"What specific information do you need?"* and wait for them to *ask* for information.
	
	   If your partner asks for information that is not on the card, do not do the calculations for them. Tell them you don't have that information.
	
	3. Before sharing the information, ask "*Why do you need that information?*"
	
	   Listen to your partner's reasoning and ask clarifying questions.
	
	4. Read the *problem card* and solve the problem independently.
	
	5. Share the *data card* and discuss your reasoning.

Pause here so your teacher can review your work. Ask your teacher for a new
set of cards and repeat the activity, trading roles with your partner.

NAME _____ DATE _____ PERIOD _____

Activity
17.3 Paper Planes

Andre, Lin, and Noah each designed and built a paper airplane. They launched each plane several times and recorded the distance of each flight in yards.

Andre	25	26	27	27	27	28	28	28	29	30	30
Lin	20	20	21	24	26	28	28	29	29	30	32
Noah	13	14	15	18	19	20	21	23	23	24	25

Work with your group to summarize the data sets with numbers and box plots.

1. Write the five-number summary for the data for each airplane.
   Then, calculate the interquartile range for each data set.

Min	Q1	Median	Q3	Max	IQR

2. Draw three box plots, one for each paper airplane. Label the box plots clearly.

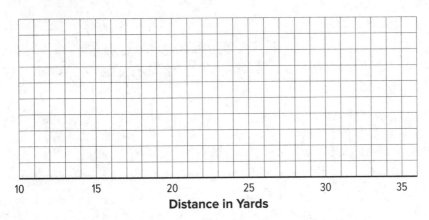

Distance in Yards

3. How are the results for Andre and Lin's planes the same? How are they different?

4. How are the results for Lin and Noah's planes the same? How are they different?

Priya joined in the paper-plane experiments. She launched her plane eleven times and recorded the lengths of each flight. She found that her maximum and minimum were equal to Lin's. Her IQR was equal to Andre's.

Draw a box plot that could represent Priya's data.

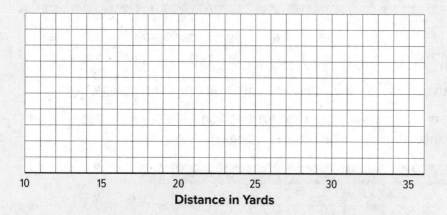

With the information given, can you estimate the median for Priya's data? Explain your reasoning.

NAME _____ DATE _____ PERIOD _____

Summary
Using Box Plots

Box plots are useful for comparing different groups. Here are two sets of plots that show the weights of some berries and some grapes.

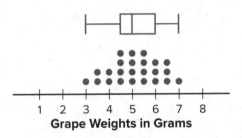

Berry Weights in Grams

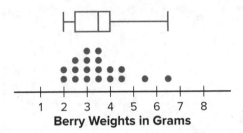

Grape Weights in Grams

Notice that the median berry weight is 3.5 grams and the median grape weight is 5 grams. In both cases, the IQR is 1.5 grams. Because the grapes in this group have a higher median weight than the berries, we can say a grape in the group is typically heavier than a berry. Because both groups have the same IQR, we can say that they have a similar variability in their weights.

These box plots represent the length data for a collection of ladybugs and a collection of beetles.

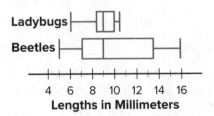

Lengths in Millimeters

The medians of the two collections are the same, but the IQR of the ladybugs is much smaller. This tells us that a typical ladybug length is similar to a typical beetle length, but the ladybugs are more alike in their length than the beetles are in their length.

1. Here are box plots that summarize the heights of 20 professional male athletes in basketball, football, hockey, and baseball.

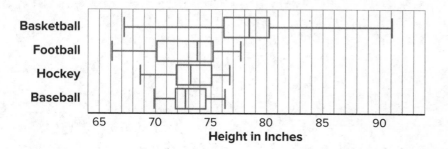

a. In which two sports are the players' height distributions most alike? Explain your reasoning.

b. Which sport shows the greatest variability in players' heights? Which sport shows the least variability?

NAME _____ DATE _____ PERIOD _____

2. Here is a box plot that summarizes data for the time, in minutes, that a fire department took to respond to 100 emergency calls. Select **all** the statements that are true, according to the box plot.

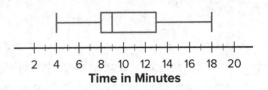

Time in Minutes

A. Most of the response times were under 13 minutes.

B. Fewer than 30 of the response times were over 13 minutes.

C. More than half of the response times were 11 minutes or greater.

D. There were more response times that were greater than 13 minutes than those that were less than 9 minutes.

E. About 75% of the response times were 13 minutes or less.

3. Pineapples were packed in three large crates. For each crate, the weight of every pineapple in the crate was recorded. Here are three box plots that summarize the weights in each crate. Select **all** of the statements that are true, according to the box plots.

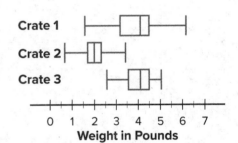

Weight in Pounds

A. The weights of the pineapples in Crate 1 were the most variable.

B. The heaviest pineapple was in Crate 1.

C. The lightest pineapple was in Crate 1.

D. Crate 3 had the greatest median weight and the greatest IQR.

E. More than half the pineapples in Crate 1 and Crate 3 were heavier than the heaviest pineapple in Crate 2.

4. Two TV shows each asked 100 viewers for their ages. For one show, the mean age of the viewers was 35 years and the MAD was 20 years. For the other show, the mean age of the viewers was 30 years and the MAD was 5 years.

A sixth-grade student says he watches one of the shows. Which show do you think he watches? Explain your reasoning. (Lesson 8-12)

Lesson 8-18

Using Data to Solve Problems

NAME _____ DATE _____ PERIOD _____

Learning Goal Let's compare data sets using visual displays.

Warm Up
18.1 Wild Bears

In one study on wild bears, researchers measured the head lengths and head widths, in inches, of 143 wild bears. The ages of the bears ranged from newborns (0 years) to 15 years. The box plots summarize the data from the study.

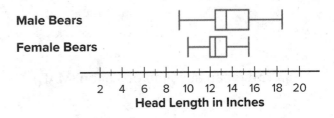

1. Write four statistical questions that could be answered using the box plots: two questions about the head length and two questions about the head width.

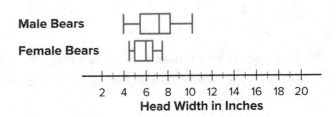

2. Trade questions with your partner.

   a. Decide if each question is a statistical question.

   b. Use the box plots to answer each question.

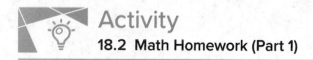

Activity

18.2 Math Homework (Part 1)

Over a two-week period, Mai recorded the number of math homework problems she had each school day.

    2   15   20   0   5   25   1   0   10   12

1. Calculate the following. Show your reasoning.

   a. The mean number of math homework problems

   b. The mean absolute deviation (MAD)

2. Interpret the mean and MAD. What do they tell you about the number of homework problems Mai had over these two weeks?

3. Find or calculate the following values and show your reasoning.

   a. The median, quartiles, maximum, and minimum of Mai's data

   b. The interquartile range (IQR)

4. Which pair of measures of center and variability—mean and MAD, or median and IQR—do you think summarizes the distribution of Mai's math homework assignments better? Explain your reasoning.

NAME _____ DATE _____ PERIOD _____

 Activity

18.3 Math Homework (Part 2)

Jada wanted to know whether a dot plot, a histogram, or a box plot would best summarize the center, variability, and other aspects of her homework data.

2   15   20   0   5   25   1   0   10   12

1. Use the axis to make a dot plot to represent the data. Mark the position of the mean, which you calculated earlier, on the dot plot using a triangle (Δ). From the triangle, draw a horizontal line segment to the left and right sides to represent the MAD.

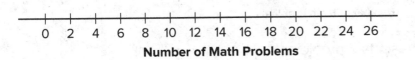

0   2   4   6   8   10   12   14   16   18   20   22   24   26

Number of Math Problems

2. Draw a box plot that represents Jada's homework data.

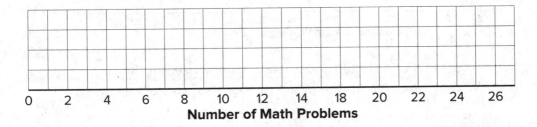

0   2   4   6   8   10   12   14   18   20   22   24   26

Number of Math Problems

3. Work with your group to draw three histograms to represent Jada's homework data. The width of the bars in each histogram should represent a different number of homework problems, which are specified as follows.

   a. The width of one bar represents 10 problems.

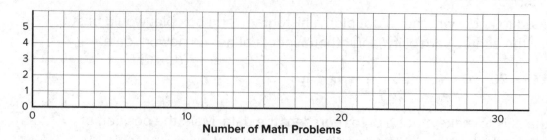

   b. The width of one bar represents 5 problems.

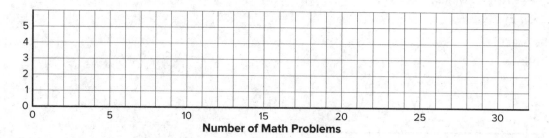

   c. The width of one bar represents 2 problems.

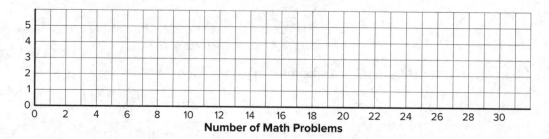

4. Which of the five representations should Jada use to summarize her data? Should she use a dot plot, box plot, or one of the histograms? Explain your reasoning.

NAME _____ DATE _____ PERIOD _____

Activity

18.4 Will the Yellow Perch Survive?

Scientists studying the yellow perch, a species of fish, believe that the length of a fish is related to its age. This means that the longer the fish, the older it is. Adult yellow perch vary in size, but they are usually between 10 and 25 centimeters.

Scientists at the Great Lakes Water Institute caught, measured, and released yellow perch at several locations in Lake Michigan. The following summary is based on a sample of yellow perch from one of these locations.

Length of Fish (centimeters)	Number of Fish
0 to less than 5	5
5 to less than 10	7
10 to less than 15	14
15 to less than 20	20
20 to less than 25	24
25 to less than 30	30

1. Use the data to make a histogram that shows the lengths of the captured yellow perch. Each bar should contain the lengths shown in each row in the table.

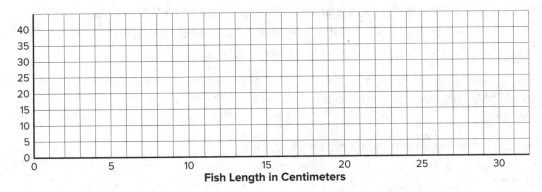

Fish Length in Centimeters

2. How many fish were measured? How do you know?

3. Use the histogram to answer the following questions.

   a. How would you describe the shape of the distribution?

   b. Estimate the median length for this sample. Describe how you made this estimate.

   c. Predict whether the mean length of this sample is greater than, less than, or nearly equal to the median length for this sample of fish? Explain your prediction.

   d. Would you use the mean or the median to describe a typical length of the fish being studied? Explain your reasoning.

4. Based on your work so far:

   a. Would you describe a typical age for the yellow perch in this sample as: "young," "adult," or "old"? Explain your reasoning.

   b. Some researchers are concerned about the survival of the yellow perch. Do you think the lengths (or the ages) of the fish in this sample are something to worry about? Explain your reasoning.

The dot plot shows the distribution of 30 cookie weights in grams.

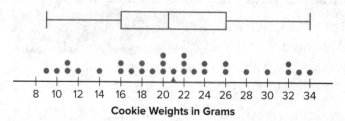

Cookie Weights in Grams

The mean cookie weight, marked by the triangle, is 21 grams. This tells us that if the weights of all of the cookies were redistributed so they all had the same weight, each cookie would weigh 21 grams. The MAD is 5.6 grams, which suggests that a cookie typically weighs between 15.4 grams and 26.6 grams.

The box plot for the same data set is shown above the dot plot. The median shows that half of the weights are greater than or equal to 20.5 grams, and half are less than or equal to 20.5 grams. The box shows that the IQR is 10 and that the middle half of the cookies weigh between 16 and 26 grams.

In this case, the median weight is very close to the mean weight, and the IQR is about twice the MAD. This tells us that the two pairs of measures of center and spread are very similar.

Now let's look at another example of 30 different cookies.

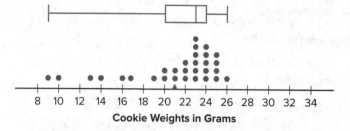

Cookie Weights in Grams

Here the mean is 21 grams, and the MAD is 3.4 grams. This suggests that a cookie typically weighs between 17.6 and 24.4 grams. The median cookie weight is 23 grams, and the box plot shows that the middle half of the data are between 20 and 24 grams. These two pairs of measures paint very different pictures of the variability of the cookie weights.

The median (23 grams) is closer to the middle of the big cluster of values. If we were to ignore the smaller cookies, the median and IQR would give a more accurate picture of how much a cookie typically weighs.

When a distribution is not symmetrical, the median and IQR are often better measures of center and spread than the mean and MAD. However, the decision on which pair of measures to use depends on what we want to know about the group we are investigating.

Learning Targets

Lesson	Learning Target(s)
8-1 Got Data?	• I can collect the correct data to answer a question and use the correct units.
	• I can explain the difference between categorical and numerical data.
8-2 Statistical Questions	• I can tell when data has variability.
8-3 Representing Data Graphically	• I can describe the information presented in tables, dot plots, and bar graphs.
	• I can use tables, dot plots, and bar graphs to represent distributions of data.
8-4 Dot Plots	• I can describe the center and spread of data from a dot plot.

(continued on the next page)

(continued from the previous page)

Lesson	Learning Target(s)
8-5 Using Dot Plots to Answer Statistical Questions	• I can use a dot plot to represent the distribution of a data set and answer questions about the real-world situation. • I can use center and spread to describe data sets, including what is typical in a data set.
8-6 Histograms	• I can recognize when a histogram is an appropriate graphical display of a data set. • I can use a histogram to get information about the distribution of data and explain what it means in a real-world situation.
8-7 Using Histograms to Answer Statistical Questions	• I can draw a histogram from a table of data. • I can use a histogram to describe the distribution of data and determine a typical value for the data.
8-8 Describing Distributions on Histograms	• I can describe the shape and features of a histogram and explain what they mean in the context of the data. • I can distinguish histograms and bar graphs.

Lesson		Learning Target(s)
8-9	Interpreting the Mean as Fair Share	• I can explain how the mean for a data set represents a "fair share." • I can find the mean for a numerical data set.
8-10	Finding and Interpreting the Mean as the Balance Point	• I can describe what the mean tells us in the context of the data. • I can explain how the mean represents a balance point for the data on a dot plot.
8-11	Deviation from the Mean	• I can find the MAD for a set of data. • I know what the mean absolute deviation (MAD) measures and what information it provides.
8-12	Using Mean and MAD to Make Comparisons	• I can say what the MAD tells us in a given context. • I can use means and MADs to compare groups.

(continued on the next page)

Lesson	Learning Target(s)
8-13 The Median of a Data Set	• I can find the median for a set of data. • I can say what the median represents and what it tells us in a given context.
8-14 Comparing Mean and Median	• I can determine when the mean or the median is more appropriate to describe the center of data. • I can explain how the distribution of data affects the mean and the median.
8-15 Quartiles and Interquartile Range	• I can use IQR to describe the spread of data. • I know what quartiles and interquartile range (IQR) measure and what they tell us about the data. • When given a list of data values or a dot plot, I can find the quartiles and interquartile range (IQR) for data.
8-16 Box Plots	• I can use the five-number summary to draw a box plot. • I know what information a box plot shows and how it is constructed.

Lesson	Learning Target(s)
8-17 Using Box Plots	• I can use a box plot to answer questions about a data set.
	• I can use medians and IQRs to compare groups.
8-18 Using Data to Solve Problems	• I can decide whether mean and MAD or median and IQR would be more appropriate for describing the center and spread of a data set.
	• I can draw an appropriate graphical representation for a set of data.
	• I can explain what the mean and MAD or the median and IQR tell us in the context of a situation and use them to answer questions.

(continued on the next page)

(continued from the previous page)

Notes:

_____

_____

_____

_____

_____

_____

_____

_____

_____

_____

_____

_____

_____

_____

_____

_____

_____

_____

_____

_____

_____

_____

Putting It All Together

In this unit, you'll apply what you've learned throughout the year to explore the mathematics of voting.

Topic
- Making Connections
- Voting

Putting It All Together

Making Connections

Voting

Lesson 9-1

Fermi Problems

NAME _____ DATE _____ PERIOD _____

Learning Goal Let's make some estimates.

Activity

1.1 Ant Trek

How long would it take an ant to run from Los Angeles to New York City?

Activity

1.2 Stacks and Stacks of Cereal Boxes

Imagine a warehouse that has a rectangular floor and that contains all of the boxes of breakfast cereal bought in the United States in one year.

If the warehouse is 10 feet tall, what could the side lengths of the floor be?

How many tiles would it take to cover
the Washington Monument?

Lesson 9-2

If Our Class Were the World

NAME _____ DATE _____ PERIOD _____

Learning Goal Let's use math to better understand our world.

 ## Activity

2.1 All 7.4 Billion of Us

There are 7.4 billion people in the world. If the whole world were represented by a 30-person class:

- 14 people would eat rice as their main food.
- 12 people would be under the age of 20.
- 5 people would be from Africa.

1. How many people in the class would *not* eat rice as their main food?

2. What percentage of the people in the class would be under the age of 20?

3. Based on the number of people in the class representing people from Africa, how many people live in Africa?

 ## Activity

2.2 About the People in the World

With the members of your group, write a list of questions about the people in the world. Your questions should begin with "How many people in the world. . ." Then, choose several questions on the list that you find most interesting.

Activity

2.3 If Our Class Were the World

Suppose your class represents all the people in the world.

Choose several characteristics about the world's population that you have investigated. Find the number of students in *your* class that would have the same characteristics.

Create a visual display that includes a diagram that represents this information. Give your display the title "If Our Class Were the World."

Lesson 9-3

Rectangle Madness

NAME _____ DATE _____ PERIOD _____

Learning Goal Let's cut up rectangles.

 Activity

3.1 Squares in Rectangles

1. Rectangle *ABCD* is not a square. Rectangle *ABEF* is a square.

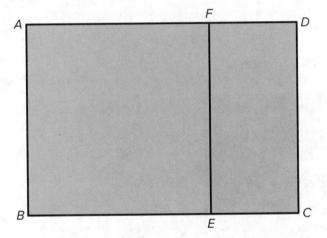

a. Suppose segment *AF* were 5 units long and segment *FD* were 2 units long. How long would segment *AD* be?

b. Suppose segment *BC* were 10 units long and segment *BE* were 6 units long. How long would segment *EC* be?

c. Suppose segment *AF* were 12 units long and segment *FD* were 5 units long. How long would segment *FE* be?

d. Suppose segment *AD* were 9 units long and segment *AB* were 5 units long. How long would segment *FD* be?

2. Rectangle *JKXW* has been decomposed into squares.

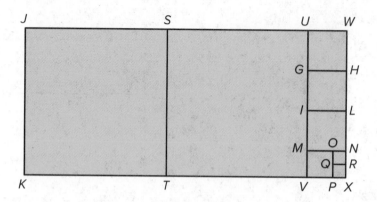

Segment *JK* is 33 units long and segment *JW* is 75 units long. Find the areas of all of the squares in the diagram.

3. Rectangle *ABCD* is 16 units by 5 units.

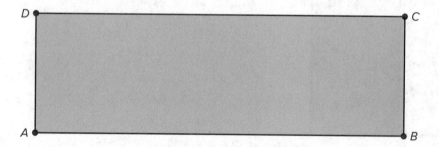

a. In the diagram, draw a line segment that decomposes *ABCD* into two regions: a square that is the largest possible and a new rectangle.

b. Draw another line segment that decomposes the *new* rectangle into two regions: a square that is the largest possible and another new rectangle.

c. Keep going until rectangle *ABCD* is entirely decomposed into squares.

d. List the side lengths of all the squares in your diagram.

NAME _____ DATE _____ PERIOD _____

Are you ready for more?

1. The diagram shows that rectangle *VWYZ* has been decomposed into three squares. What could the side lengths of this rectangle be?

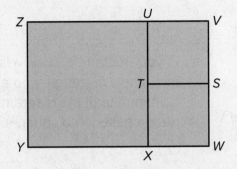

2. How many different side lengths can you find for rectangle *VWYZ*?

3. What are some rules for possible side lengths of rectangle *VWYZ*?

 ## Activity

3.2 More Rectangles, More Squares

1. Draw a rectangle that is 21 units by 6 units.

   a. In your rectangle, draw a line segment that decomposes the rectangle into a new rectangle and a square that is as large as possible. Continue until the diagram shows that your original rectangle has been entirely decomposed into squares.

   b. How many squares of each size are in your diagram?

   c. What is the side length of the smallest square?

2. Draw a rectangle that is 28 units by 12 units.

a. In your rectangle, draw a line segment that decomposes the rectangle into a new rectangle and a square that is as large as possible. Continue until the diagram shows that your original rectangle has been decomposed into squares.

b. How many squares of each size are in your diagram?

c. What is the side length of the smallest square?

3. Write each of these fractions as a mixed number with the smallest possible numerator and denominator:

a. $\dfrac{16}{5}$

b. $\dfrac{21}{6}$

c. $\dfrac{28}{12}$

4. What do the fraction problems have to do with the previous rectangle decomposition problems?

NAME _____ DATE _____ PERIOD _____

Activity
3.3 Finding Equivalent Fractions

1. Accurately draw a rectangle that is 9 units by 4 units.

   a. In your rectangle, draw a line segment that decomposes the rectangle into a new rectangle and a square that is as large as possible. Continue until your original rectangle has been entirely decomposed into squares.

   b. How many squares of each size are there?

   c. What are the side lengths of the last square you drew?

   d. Write $\frac{9}{4}$ as a mixed number.

2. Accurately draw a rectangle that is 27 units by 12 units.

   a. In your rectangle, draw a line segment that decomposes the rectangle into a new rectangle and a square that is as large as possible. Continue until your original rectangle has been entirely decomposed into squares.

   b. How many squares of each size are there?

   c. What are the side lengths of the last square you drew?

   d. Write $\frac{27}{12}$ as a mixed number.

   e. Compare the diagram you drew for this problem and the one for the previous problem. How are they the same? How are they different?

3. What is the greatest common factor of 9 and 4? What is the greatest common factor of 27 and 12? What does this have to do with your diagrams of decomposed rectangles?

Are you ready for more?

We have seen some examples of rectangle tilings. A *tiling* means a way to completely cover a shape with other shapes, without any gaps or overlaps. For example, here is a tiling of rectangle *KXWJ* with 2 large squares, 3 medium squares, 1 small square, and 2 tiny squares.

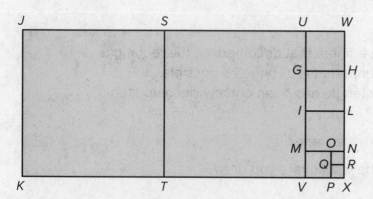

Some of the squares used to tile this rectangle have the same size.

Might it be possible to tile a rectangle with squares where the squares are *all different sizes*?

If you think it is possible, find such a rectangle and such a tiling.
If you think it is not possible, explain why it is not possible.

NAME _____ DATE _____ PERIOD _____

 ## Activity
3.4 It's All About Fractions

1. Accurately draw a 37-by-16 rectangle.
   (Use graph paper, if possible.)

   a. In your rectangle, draw a line segment that
      decomposes the rectangle into a
      new rectangle and a square that is as large
      as possible. Continue until your original
      rectangle has been entirely decomposed
      into squares.

   b. How many squares of each size are there?

   c. What are the dimensions of the last square
      you drew?

   d. What does this have to do with $2 + \dfrac{1}{3 + \frac{1}{5}}$?

2. Consider a 52-by-15 rectangle.

   a. In your rectangle, draw a line segment that
      decomposes the rectangle into a
      new rectangle and a square that is as large
      as possible. Continue until your original
      rectangle has been entirely decomposed
      into squares.

   b. Write a fraction equal to this expression:
      $3 + \dfrac{1}{2 + \frac{1}{7}}$.

   c. Notice some connections between the
      rectangle and the fraction.

   d. What is the greatest common factor of
      52 and 15?

3. Consider a 98-by-21 rectangle.

   a. In your rectangle, draw a line segment that decomposes the rectangle into a new rectangle and a square that is as large as possible. Continue until your original rectangle has been entirely decomposed into squares.

   b. Write a fraction equal to this expression:
   $$4 + \frac{1}{1 + \frac{7}{14}}.$$

   c. Notice some connections between the rectangle and the fraction.

   d. What is the greatest common factor of 98 and 21?

4. Consider a 121-by-38 rectangle.

   a. Use the decomposition-into-squares process to write a continued fraction for $\frac{121}{38}$. Verify that it works.

   b. What is the greatest common factor of 121 and 38?

Lesson 9-4

How Do We Choose?

NAME _____ DATE _____ PERIOD _____

Learning Goal Let's vote and choose a winner!

Activity

4.1 Which Was "Yessier"?

Two sixth-grade classes, A and B, voted on whether to give the answers to their math problems in poetry. The "yes" choice was more popular in both classes.

Was one class more in favor of math poetry, or were they equally in favor? Find three or more ways to answer the question.

	Yes	No
Class A	24	16
Class B	18	9

Activity

4.2 Which Class Voted Purpler?

The school will be painted over the summer. Students get to vote on whether to change the color to purple (a "yes" vote) or keep it a beige color (a "no" vote). The principal of the school decided to analyze voting results by class. The table shows some results.

	Yes	No
Class A	26	14
Class B	31	19

In both classes, a majority voted for changing the paint color to purple. Which class was more in favor of changing?

Activity

4.3 Supermajorities

1. Another school is also voting on whether to change their school's color to purple. Their rules require a $\frac{2}{3}$ supermajority to change the colors. A total of 240 people voted, and 153 voted to change to purple. Were there enough votes to make the change?

2. This school also is thinking of changing their mascot to an armadillo. To change mascots, a 55% supermajority is needed. How many of the 240 students need to vote "yes" for the mascot to change?

3. At this school, which requires more votes to pass: a change of mascot or a change of color?

Activity

4.4 Best Restaurant

A town's newspaper held a contest to decide the best restaurant in town. Only people who subscribe to the newspaper can vote. 25% of the people in town subscribe to the newspaper. 20% of the subscribers voted. 80% of the people who voted liked Darnell's BBQ Pit best.

Darnell put a big sign in his restaurant's window that said, "80% say Darnell's is the best!"

Do you think Darnell's sign is making an accurate statement? Support your answer with:

- Some calculations

- An explanation in words

- A diagram that accurately represents the people in town, the newspaper subscribers, the voters, and the people who liked Darnell's best

Lesson 9-5

More Than Two Choices

NAME _____ DATE _____ PERIOD _____

Learning Goal Let's explore different ways to determine a winner.

 **Activity**

5.1 Field Day

Students in a sixth-grade class were asked, "What activity would you most like to do for field day?" The results are shown in the table.

Activity	Number of Votes
Softball Game	16
Scavenger Hunt	10
Dancing Talent Show	8
Marshmallow Throw	4
No Preference	2

1. What percentage of the class voted for softball?

2. What percentage did not vote for softball as their first choice?

Activity

5.2 School Lunches (Part 1)

Suppose students at our school are voting for the lunch menu over the course of one week. The following is a list of options provided by the caterer.

1. Meat Lovers
   - Meat loaf
   - Hot dogs
   - Pork cutlets
   - Beef stew
   - Liver and onions

2. Vegetarian
   - Vegetable soup and peanut butter sandwich
   - Hummus, pita, and veggie sticks
   - Veggie burgers and fries
   - Chef's salad
   - Cheese pizza every day
   - Double desserts every day

3. Something for Everyone
   - Chicken nuggets
   - Burgers and fries
   - Pizza
   - Tacos
   - Leftover day (all the week's leftovers made into a casserole)
   - Bonus side dish: pea jello (green gelatin with canned peas)

4. Concession Stand
   - Choice of hamburger or hot dog, with fries, every day

NAME _____ DATE _____ PERIOD _____

To vote, draw one of the following symbols next to each menu option to show your first, second, third, and last choices. If you use the slips of paper from your teacher, use only the column that says "symbol."

    1st choice   2nd choice   3rd choice   4th choice

1. Meat Lovers _____

2. Vegetarian _____

3. Something for Everyone _____

4. Concession Stand _____

Here are two voting systems that can be used to determine the winner.

Voting System #1. *Plurality*: The option with the most first-choice votes (stars) wins.

Voting System #2. *Runoff*: If no choice received a majority of the votes, leave out the choice that received the fewest first-choice votes (stars). Then have another vote.

If your first vote is still a choice, vote for that. If not, vote for your second choice that you wrote down.

If there is still no majority, leave out the choice that got the fewest votes, and then vote again. Vote for your first choice if it's still in, and if not, vote for your second choice. If your second choice is also out, vote for your third choice.

1. How many people in our class are voting? How many votes does it take to win a majority?

2. How many votes did the top option receive? Was this a majority of the votes?

3. People tend to be more satisfied with election results if their top choices win. For how many, and what percentage, of people was the winning option:

   a. their first choice?

   b. their second choice?

   c. their third choice?

   d. their last choice?

4. After the second round of voting, did any choice get a majority? If so, is it the same choice that got a plurality in Voting System #1?

5. Which choice won?

6. How satisfied were the voters by the election results? For how many, and what percentage, of people was the winning option:

   a. their first choice?

   b. their second choice?

   c. their third choice?

   d. their last choice?

7. Compare the satisfaction results for the plurality voting rule and the runoff rule. Did one produce satisfactory results for more people than the other?

NAME _____ DATE _____ PERIOD _____

Activity
5.3 School Lunch (Part 2)

Let's analyze a different election.

In another class, there are four clubs. Everyone in each club agrees to vote for the lunch menu exactly the same way, as shown in this table.

	Barbecue Club (21 members)	Garden Club (13 members)	Sports Boosters (7 members)	Film Club (9 members)
A. Meat Lovers	★	✕	✕	✕
B. Vegetarian	☺	★	😐	😐
C. Something for Everyone	😐	😐	☺	★
D. Concession Stand	✕	☺	★	☺

1. Figure out which option won the election by answering these questions.

   a. On the first vote, when everyone voted for their first choice, how many votes did each option get? Did any choice get a majority?

   b. Which option is removed from the next vote?

   c. On the second vote, how many votes did each of the remaining three menu options get? Did any option get a majority?

   d. Which menu option is removed from the next vote?

   e. On the third vote, how many votes did each of the remaining two options get? Which option won?

2. Estimate how satisfied all the voters were.

   a. For how many people was the winner their first choice?

   b. For how many people was the winner their second choice?

   c. For how many people was the winner their third choice?

   d. For how many people was the winner their last choice?

3. Compare the satisfaction results for the plurality voting rule and the runoff rule. Did one produce satisfactory results for more people than the other?

Activity

5.4 Just Vote Once

Your class just voted using the *instant runoff* system. Use the class data for following questions.

1. For our class, which choice received the most points?

2. Does this result agree with that from the runoff election in an earlier activity?

3. For the other class, which choice received the most points?

4. Does this result agree with that from the runoff election in an earlier activity?

5. The runoff method uses information about people's first, second, third, and last choices when it is not clear that there is a winner from everyone's first choices. How does the instant runoff method include the same information?

6. After comparing the results for the three voting rules (plurality, runoff, instant runoff) and the satisfaction surveys, which method do you think is fairest? Explain.

NAME _____ DATE _____ PERIOD _____

Are you ready for more?

Numbering your choices 0 through 3 might not really describe your opinions. For example, what if you really liked A and C a lot, and you really hated B and D? You might want to give A and C both a 3, and B and D both a 0.

1. Design a numbering system where the size of the number accurately shows how much you like a choice. Some ideas:

   - The same 0 to 3 scale, but you can choose more than one of each number, or even decimals between 0 and 3.

   - A scale of 1 to 10, with 10 for the best and 1 for the worst.

2. Try out your system with the people in your group, using the same school lunch options for the election.

3. Do you think your system gives a more fair way to make choices? Explain your reasoning.

Activity
5.5 Weekend Choices

Clare, Han, Mai, Tyler, and Noah are deciding what to do on the weekend. Their options are cooking, hiking, and bowling. Here are the points for their instant runoff vote. Each first choice gets 2 points, the second choice gets 1 point, and the last choice gets 0 points.

	Cooking	Hiking	Bowling
Clare	2	1	0
Han	2	1	0
Mai	2	1	0
Tyler	0	2	1
Noah	0	2	1

1. Which activity won using the instant runoff method? Show your calculations and use expressions or equations.

2. Which activity would have won if there was just a vote for their top choice, with a majority or plurality winning?

3. Which activity would have won if there was a runoff election?

4. Explain why this happened.

Lesson 9-6

Picking Representatives

NAME _____ DATE _____ PERIOD _____

Learning Goal Let's think about fair representation.

 ## Activity

6.1 Computers for Kids

A program gives computers to families with school-aged children. They have a certain number of computers to distribute fairly between several families. How many computers should each family get?

1. One month the program has 8 computers. The families have these numbers of school-aged children: 4, 2, 6, 2, 2.

   a. How many children are there in all?

   b. Counting all the children in all the families, how many children would use each computer? This is the number of children per computer. Call this number *A*.

   c. Fill in the third column of the table. Decide how many computers to give to each family if we use *A* as the basis for distributing the computers.

Family	Number of Children	Number of Computers, Using *A*
Baum	4	
Chu	2	
Davila	6	
Eno	2	
Farouz	2	

   d. Check that 8 computers have been given out in all.

2. The next month they again have 8 computers. There are different families with these numbers of children: 3, 1, 2, 5, 1, 8.

   a. How many children are there in all?

   b. Counting all the children in all the families, how many children would use each computer? This is the number of children per computer. Call this number B.

   c. Does it make sense that B is not a whole number? Why?

   d. Fill in the third column of the table. Decide how many computers to give to each family if we use B as the basis for distributing the computers.

Family	Number of Children	Number of Computers, Using B	Number of Computers, Your Way	Number of Children per Computer, Your Way
Gray	3			
Hernandez	1			
Ito	2			
Jones	5			
Krantz	1			
Lo	8			

   e. Check that 8 computers have been given out in all.

   f. Does it make sense that the number of computers for one family is not a whole number? Explain your reasoning?

   g. Find and describe a way to distribute computers to the families so that each family gets a whole number of computers. Fill in the fourth column of the table.

   h. Compute the number of children per computer in each family and fill in the last column of the table.

   i. Do you think your way of distributing the computers is fair? Explain your reasoning.

NAME _____ DATE _____ PERIOD _____

Activity
6.2 School Mascot (Part 1)

A school is deciding on a school mascot. They have narrowed the choices down to the Banana Slugs or the Sea Lions.

The principal decided that each class gets one vote. Each class held an election, and the winning choice was the one vote for the whole class. The table shows how three classes voted.

	Banana Slugs	Sea Lions	Class Vote
Class A	9	3	banana slug
Class B	14	10	
Class C	6	30	

1. Which mascot won, according to the principal's plan? What percentage of the votes did the winner get under this plan?

2. Which mascot received the most student votes in all? What percentage of the votes did this mascot receive?

3. The students thought this plan was not very fair. They suggested that bigger classes should have more votes to send to the principal.
   Make up a proposal for the principal where there are as few votes as possible, but the votes proportionally represent the number of students in each class.

4. Decide how to assign the votes for the results in the class. (Do they all go to the winner? Or should the loser still get some votes?)

5. In your system, which mascot is the winner?

6. In your system, how many representative votes are there? How many students does each vote represent?

1. In a very small school district, there are four schools, D, E, F, and G. The district wants a total of 10 advisors for the students. Each school should have at least one advisor.

School	Number of Students	Number of Advisors, *A* Students per Advisor
D	48	
E	12	
F	24	
G	36	

a. How many students are in this district in all?

b. If the advisors could represent students at different schools, how many students per advisor should there be? Call this number *A*.

c. Using *A* students per advisor, how many advisors should each school have? Complete the table with this information for schools D, E, F, and G.

NAME _____ DATE _____ PERIOD _____

2. Another district has four schools; some are large, others are small. The district wants 10 advisors in all. Each school should have at least one advisor.

School	Number of Students	Number of Advisors, B Students per Advisor	Number of Advisors, Your Way	Number of Students per Advisor, Your Way
Dr. King School	500			
O'Connor School	200			
Science Magnet School	140			
Trombone Academy	10			

a. How many students are in this district in all?

b. If the advisors didn't have to represent students at the same school, how many students per advisor should there be? Call this number B.

c. Using B students per advisor, how many advisors should each school have? Give your quotients to the tenths place. Fill in the first "number of advisors" column of the table. Does it make sense to have a tenth of an advisor?

d. Decide on a consistent way to assign advisors to schools so that there are only whole numbers of advisors for each school, and there is a total of 10 advisors among the schools. Fill in the "your way" column of the table.

e. How many students per advisor are there at each school? Fill in the last row of the table.

f. Do you think this is a fair way to assign advisors? Explain your reasoning.

Activity
6.4 School Mascot (Part 2)

The whole town gets interested in choosing a mascot. The mayor of the town decides to choose representatives to vote.

There are 50 blocks in the town, and the people on each block tend to have the same opinion about which mascot is best. Green blocks like sea lions, and gold blocks like banana slugs. The mayor decides to have 5 representatives, each representing a district of 10 blocks.

Here is a map of the town, with preferences shown.

1. Suppose there were an election with each block getting one vote. How many votes would be for banana slugs? For sea lions? What percentage of the vote would be for banana slugs?

2. Suppose the districts are shown in the next map. What did the people in each district prefer? What did their representative vote? Which mascot would win the election?

Complete the table with this election's results.

District	Number of Blocks for Banana Slugs	Number of Blocks for Sea Lions	Percentage of Blocks for Banana Slugs	Representative's Vote
1	10	0		banana slugs
2				
3				
4				
5				

3. Suppose, instead, that the districts are shown in the new map below. What did the people in each district prefer? What did their representative vote? Which mascot would win the election?

Complete the table with this election's results.

District	Number of Blocks for Banana Slugs	Number of Blocks for Sea Lions	Percentage of Blocks for Banana Slugs	Representative's Vote
1				
2				
3				
4				
5				

4. Suppose the districts are designed in yet another way, as shown in the next map. What did the people in each district prefer? What did their representative vote? Which mascot would win the election?

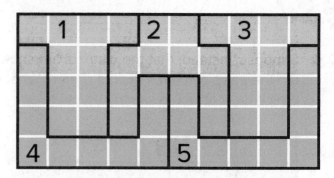

Complete the table with this election's results.

District	Number of Blocks for Banana Slugs	Number of Blocks for Sea Lions	Percentage of Blocks for Banana Slugs	Representative's Vote
1				
2				
3				
4				
5				

5. Write a headline for the local newspaper for each of the ways of splitting the town into districts.

6. Which systems on the three maps of districts do you think are more fair? Are any totally unfair?

NAME _____ DATE _____ PERIOD _____

Activity

6.5 Fair and Unfair Districts

1. Smallville's map is shown, with opinions shown by block in green and gold. Decompose the map to create three connected, equal-area districts in two ways:

a. Design three districts where *green* will win at least two of the three districts. Record results in Table 1.

Table 1:

District	Number of Blocks for Green	Number of Blocks for Gold	Percentage of Blocks for Green	Representative's Vote
1				
2				
3				

b. Design three districts where *gold* will win at least two of the three districts. Record results in Table 2.

Table 2:

District	Number of Blocks for Green	Number of Blocks for Gold	Percentage of Blocks for Green	Representative's Vote
1				
2				
3				

2. Squaretown's map is shown, with opinions by block shown in green and gold. Decompose the map to create five connected, equal-area districts in two ways:

a. Design five districts where *green* will win at least three of the five districts. Record the results in Table 3.

Table 3:

District	Number of Blocks for Green	Number of Blocks for Gold	Percentage of Blocks for Green	Representative's Vote
1				
2				
3				
4				
5				

b. Design five districts where *gold* will win at least three of the five districts. Record the results in Table 4.

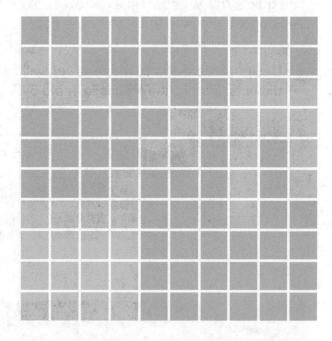

Table 4:

District	Number of Blocks for Green	Number of Blocks for Gold	Percentage of Blocks for Green	Representative's Vote
1				
2				
3				
4				
5				

3. Mountain Valley's map is shown, with opinions by block shown in green and gold. (This is a town in a narrow valley in the mountains.) Can you decompose the map to create three connected, equal-area districts in the two ways described here?

a. Design three districts where *green* will win at least two of the three districts. Record the results in Table 5.

Table 5:

District	Number of Blocks for Green	Number of Blocks for Gold	Percentage of Blocks for Green	Representative's Vote
1				
2				
3				

b. Design three districts where *gold* will win at least two of the three districts. Record the results in Table 6.

Table 6:

District	Number of Blocks for Green	Number of Blocks for Gold	Percentage of Blocks for Green	Representative's Vote
1				
2				
3				

Glossary

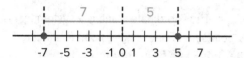

A

absolute value The absolute value of a number is its distance from 0 on the number line.

The absolute value of -7 is 7, because it is 7 units away from 0. The absolute value of 5 is 5, because it is 5 units away from 0.

area Area is the number of square units that cover a two-dimensional region, without any gaps or overlaps.

For example, the area of region A is 8 square units. The area of the shaded region of B is $\frac{1}{2}$ square unit.

Region A   **Region B**

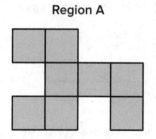

average The average is another name for the mean of a data set.

For the data set 3, 5, 6, 8, 11, 12, the average is 7.5.

$3 + 5 + 6 + 8 + 11 + 12 = 45$

$45 \div 6 = 7.5$

B

base (of a parallelogram or triangle) We can choose any side of a parallelogram or triangle to be the shape's base. Sometimes we use the word *base* to refer to the length of this side.

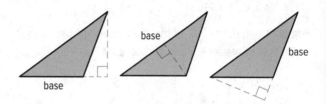

base (of a prism or pyramid) The word *base* can also refer to a face of a polyhedron.

A prism has two identical bases that are parallel. A pyramid has one base.

A prism or pyramid is named for the shape of its base.

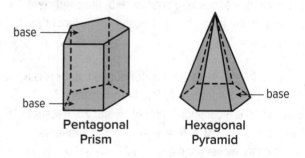

Pentagonal
Prism

Hexagonal
Pyramid

box plot A box plot is a way to represent data on a number line. The data is divided into four sections. The sides of the box represent the first and third quartiles. A line inside the box represents the median. Lines outside the box connect to the minimum and maximum values.

For example, this box plot shows a data set with a minimum of 2 and a maximum of 15. The median is 6, the first quartile is 5, and the third quartile is 10.

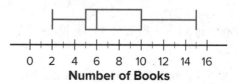

Number of Books

C

categorical data A set of categorical data has values that are words instead of numbers.

For example, Han asks 5 friends to name their favorite color. Their answers are: blue, blue, green, blue, orange.

center The center of a set of numerical data is a value in the middle of the distribution. It represents a typical value for the data set.

For example, the center of this distribution of cat weights is between 4.5 and 5 kilograms.

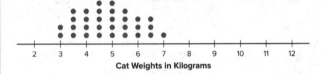

Cat Weights in Kilograms

coefficient A coefficient is a number that is multiplied by a variable.

For example, in the expression $3x + 5$, the coefficient of x is 3. In the expression $y + 5$, the coefficient of y is 1, because $y = 1 \cdot y$.

common factor A common factor of two numbers is a number that divides evenly into both numbers. For example, 5 is a common factor of 15 and 20, because $15 \div 5 = 3$ and $20 \div 5 = 4$. Both of the quotients, 3 and 4, are whole numbers.

- The factors of 15 are 1, 3, *5*, and 15.

- The factors of 20 are 1, 2, 4, *5*, 10, and 20.

common multiple A common multiple of two numbers is a product you can get by multiplying each of the two numbers by some whole number. For example, 30 is a common multiple of 3 and 5, because $3 \cdot 10 = 30$ and $5 \cdot 6 = 30$. Both of the factors, 10 and 6, are whole numbers.

- The multiples of 3 are 3, 6, 9, 12, *15*, 18, 21, 24, 27, *30*, 33 . . .

- The multiples of 5 are 5, 10, *15*, 20, 25, *30*, 35, 40 . . .

compose Compose means "put together." We use the word *compose* to describe putting more than one figure together to make a new shape.

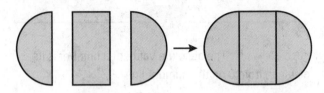

coordinate plane The coordinate plane is a system for telling where points are. For example, point R is located at $(3, 2)$ on the coordinate plane, because it is three units to the right and two units up.

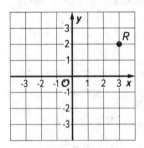

cubed We use the word *cubed* to mean "to the third power." This is because a cube with side length s has a volume of $s \cdot s \cdot s$, or s^3.

D

decompose Decompose means "take apart." We use the word *decompose* to describe taking a figure apart to make more than one new shape.

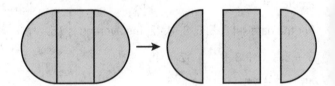

dependent variable The dependent variable is the result of a calculation.

For example, a boat travels at a constant speed of 25 miles per hour. The equation $d = 25t$ describes the relationship between the boat's distance and time. The dependent variable is the distance traveled, because d is the result of multiplying 25 by t.

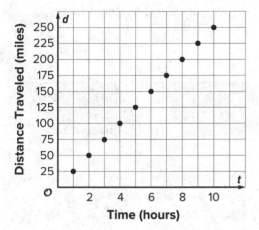

distribution The distribution tells how many times each value occurs in a data set. For example, in the data set blue, blue, green, blue, orange, the distribution is 3 blues, 1 green, and 1 orange.

Here is a dot plot that shows the distribution for the data set 6, 10, 7, 35, 7, 36, 32, 10, 7, 35.

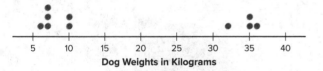

Dog Weights in Kilograms

dot plot A dot plot is a way to represent data on a number line. Each time a value appears in the data set, we put another dot above that number on the number line.

For example, in this dot plot there are three dots above the 9. This means that three different plants had a height of 9 cm.

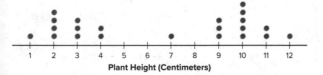

Plant Height (Centimeters)

double number line diagram A double number line diagram uses a pair of parallel number lines to represent equivalent ratios. The locations of the tick marks match on both number lines. The tick marks labeled 0 line up, but the other numbers are usually different.

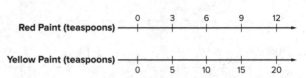

E

edge Each straight side of a polygon is called an edge. For example, the edges of this polygon are segments AB, BC, CD, DE, and EA.

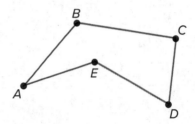

equivalent expressions Equivalent expressions are always equal to each other. If the expressions have variables, they are equal whenever the same value is used for the variable in each expression.

For example, $3x + 4x$ is equivalent to $5x + 2x$. No matter what value we use for x, these expressions are always equal. When x is 3, both expressions equal 21. When x is 10, both expressions equal 70.

equivalent ratios Two ratios are equivalent if you can multiply each of the numbers in the first ratio by the same factor to get the numbers in the second ratio.

For example, 8 : 6 is equivalent to 4 : 3, because $8 \cdot \frac{1}{2} = 4$ and $6 \cdot \frac{1}{2} = 3$.

A recipe for lemonade says to use 8 cups of water and 6 lemons. If we use 4 cups of water and 3 lemons, it will make half as much lemonade. Both recipes taste the same, because 8 : 6 and 4 : 3 are equivalent ratios.

Cups of Water	Number of Lemons
8	6
4	3

exponent In expressions like 5^3 and 8^2, the 3 and the 2 are called exponents. They tell you how many factors to multiply. For example, $5^3 = 5 \cdot 5 \cdot 5$, and $8^2 = 8 \cdot 8$.

face Each flat side of a polyhedron is called a face. For example, a cube has 6 faces, and they are all squares.

frequency The frequency of a data value is how many times it occurs in the data set.

For example, there were 20 dogs in a park. The table shows the frequency of each color.

Color	Frequency
White	4
Brown	7
Black	3
Multi-Color	6

greatest common factor The greatest common factor of two numbers is the largest number that divides evenly into both numbers. Sometimes we call this the GCF. For example, 15 is the greatest common factor of 45 and 60.

- The factors of 45 are 1, 3, 5, 9, *15*, and 45.
- The factors of 60 are 1, 2, 3, 4, 5, 6, 10, 12, *15*, 20, 30, and 60.

height (of a parallelogram or triangle) The height is the shortest distance from the base of the shape to the opposite side (for a parallelogram) or opposite vertex (for a triangle).

We can show the height in more than one place, but it will always be perpendicular to the chosen base.

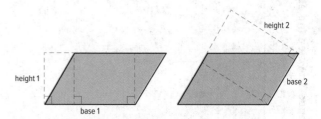

histogram A histogram is a way to represent data on a number line. Data values are grouped by ranges. The height of the bar shows how many data values are in that group.

This histogram shows there were 10 people who earned 2 or 3 tickets. We can't tell how many of them earned 2 tickets or how many earned 3. Each bar includes the left-end value but not the right-end value. (There were 5 people who earned 0 or 1 tickets and 13 people who earned 6 or 7 tickets.)

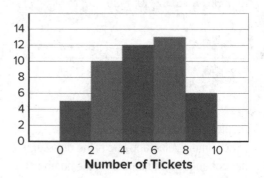

I

independent variable The independent variable is used to calculate the value of another variable.

For example, a boat travels at a constant speed of 25 miles per hour. The equation $d = 25t$ describes the relationship between the boat's distance and time. The independent variable is time, because t is multiplied by 25 to get d.

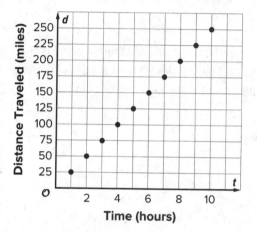

interquartile range (IQR) The interquartile range is one way to measure how spread out a data set is. We sometimes call this the IQR. To find the interquartile range we subtract the first quartile from the third quartile.

For example, the IQR of this data set is 20 because $5 - 30 = 20$.

22	29	30	31	32	43	44	45	50	50	59
		Q1			Q2			Q3		

L

least common multiple The least common multiple of two numbers is the smallest product you can get by multiplying each of the two numbers by some whole number. Sometimes we call this the LCM. For example, 30 is the least common multiple of 6 and 10.

- The multiples of 6 are 6, 12, 18, 24, *30*, 36, 42, 48, 54, 60 . . .
- The multiples of 10 are 10, 20, *30*, 40, 50, 60, 70, 80 . . .

long division Long division is a way to show the steps for dividing numbers in decimal form. It finds the quotient one digit at a time, from left to right. For example, here is the long division for $57 \div 4$.

```
      14.25
  4 ) 57.00
     -4
     ___
      17
     -16
     ___
       10
       -8
      ___
        20
       -20
       ___
         0
```

M

mean The mean is one way to measure the center of a data set. We can think of it as a balance point. For example, for the data set 7, 9, 12, 13, 14, the mean is 11.

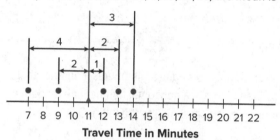
Travel Time in Minutes

To find the mean, add up all the numbers in the data set. Then, divide by how many numbers there are. $7 + 9 + 12 + 13 + 14 = 55$ and $55 \div 5 = 11$.

mean absolute deviation (MAD) The mean absolute deviation is one way to measure how spread out a data set is. Sometimes we call this the MAD. For example, for the data set 7, 9, 12, 13, 14, the MAD is 2.4. This tells us that these travel times are typically 2.4 minutes away from the mean, which is 11.

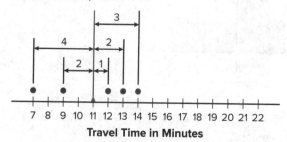
Travel Time in Minutes

To find the MAD, add up the distance between each data point and the mean. Then, divide by how many numbers there are. $4 + 2 + 1 + 2 + 3 = 12$ and $12 \div 5 = 2.4$.

measure of center A measure of center is a value that seems typical for a data distribution.

Mean and median are both measures of center.

median The median is one way to measure the center of a data set. It is the middle number when the data set is listed in order.

For the data set 7, 9, 12, 13, 14, the median is 12.

For the data set 3, 5, 6, 8, 11, 12, there are two numbers in the middle. The median is the average of these two numbers. $6 + 8 = 14$ and $14 \div 2 = 7$.

meters per second Meters per second is a unit for measuring speed. It tells how many meters an object goes in one second.

For example, a person walking 3 meters per second is going faster than another person walking 2 meters per second.

negative number A negative number is a number that is less than zero. On a horizontal number line, negative numbers are usually shown to the left of 0.

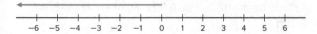

net A net is a two-dimensional figure that can be folded to make a polyhedron.

Here is a net for a cube.

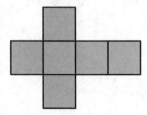

numerical data A set of numerical data has values that are numbers.

For example, Han lists the ages of people in his family: 7, 10, 12, 36, 40, 67.

opposite Two numbers are opposites if they are the same distance from 0 and on different sides of the number line.

For example, 4 is the opposite of -4, and -4 is the opposite of 4. They are both the same distance from 0. One is negative, and the other is positive.

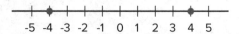

opposite vertex For each side of a triangle, there is one vertex that is not on that side. This is the opposite vertex. For example point *A* is the opposite vertex to side *BC*.

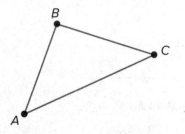

pace Pace is one way to describe how fast something is moving. Pace tells how much time it takes the object to travel a certain distance.

For example, Diego walks at a pace of 10 minutes per mile. Elena walks at a pace of 11 minutes per mile. Elena walks slower than Diego, because it takes her more time to travel the same distance.

parallelogram A parallelogram is a type of quadrilateral that has two pairs of parallel sides.

Here are two examples of parallelograms.

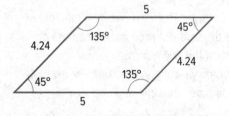

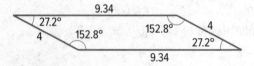

per The word *per* means "for each." For example, if the price is $5 per ticket, that means you will pay $5 *for each* ticket. Buying 4 tickets would cost $20, because $4 \cdot 5 = 20$.

percent The word *percent* means "for each 100." The symbol for percent is %.

For example, a quarter is worth 25 cents, and a dollar is worth 100 cents. We can say that a quarter is worth 25% of a dollar.

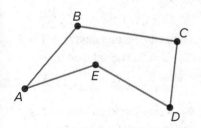

| 1 Quarter | 25¢ |
| 1 Dollar | 100¢ |

percentage A percentage is a rate per 100.

For example, a fish tank can hold 36 liters. Right now there are 27 liters of water in the tank. The percentage of the tank that is full is 75%.

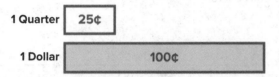

polygon A polygon is a closed, two-dimensional shape with straight sides that do not cross each other.

Figure *ABCDE* is an example of a polygon.

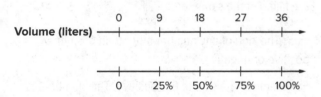

polyhedron (polyhedra) A polyhedron is a closed, three-dimensional shape with flat sides. When we have more than one polyhedron, we call them polyhedra.

Here are some drawings of polyhedra.

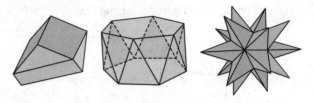

positive number A positive number is a number that is greater than zero. On a horizontal number line, positive numbers are usually shown to the right of 0.

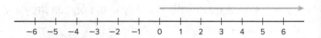

prism A prism is a type of polyhedron that has two bases that are identical copies of each other. The bases are connected by rectangles or parallelograms.

Here are some drawings of some prisms.

  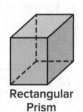

Triangular Prism   Pentagonal Prism   Rectangular Prism

pyramid A pyramid is a type of polyhedron that has one base. All the other faces are triangles, and they all meet at a single vertex

Here are some drawings of pyramids.

Rectangular Pyramid   Hexagonal Pyramid   Heptagonal Pyramid

Q

quadrant The coordinate plane is divided into 4 regions called quadrants. The quadrants are numbered using Roman numerals, starting in the top right corner.

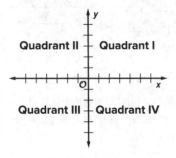

quadrilateral A quadrilateral is a type of polygon that has 4 sides. A rectangle is an example of a quadrilateral. A pentagon is not a quadrilateral, because it has 5 sides.

quartile Quartiles are the numbers that divide a data set into four sections that each have the same number of values.

For example, in this data set the first quartile is 30. The second quartile is the same thing as the median, which is 43. The third quartile is 50.

22	29	30	31	32	43	44	45	50	50	59
		Q1			Q2			Q3		

R

range The range is the distance between the smallest and largest values in a data set. For example, for the data set 3, 5, 6, 8, 11, 12, the range is 9, because $12 - 3 = 9$.

ratio A ratio is an association between two or more quantities.

For example, the ratio 3 : 2 could describe a recipe that uses 3 cups of flour for every 2 eggs, or a boat that moves 3 meters every 2 seconds. One way to represent the ratio 3 : 2 is with a diagram that has 3 blue squares for every 2 green squares.

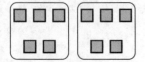

rational number A rational number is a fraction or the opposite of a fraction.

For example, 8 and -8 are rational numbers because they can be written as $\frac{8}{1}$ and, $-\frac{8}{1}$.

Also, 0.75 and -0.75 are rational numbers because they can be written as $\frac{75}{100}$ and, $-\frac{75}{100}$.

reciprocal Dividing 1 by a number gives the reciprocal of that number. For example, the reciprocal of 12 is $\frac{1}{12}$, and the reciprocal of $\frac{2}{5}$ is $\frac{5}{2}$.

region A region is the space inside of a shape. Some examples of two-dimensional regions are inside a circle or inside a polygon. Some examples of three-dimensional regions are the inside of a cube or the inside of a sphere.

S

same rate We use the words *same rate* to describe two situations that have equivalent ratios.

For example, a sink is filling with water at a rate of 2 gallons per minute. If a tub is also filling with water at a rate of 2 gallons per minute, then the sink and the tub are filling at the same rate.

sign The sign of any number other than 0 is either positive or negative.

For example, the sign of 6 is positive. The sign of -6 is negative. Zero does not have a sign, because it is not positive or negative.

solution to an equation A solution to an equation is a number that can be used in place of the variable to make the equation true.

For example, 7 is the solution to the equation $m + 1 = 8$, because it is true that $7 + 1 = 8$. The solution to $m + 1 = 8$ is not 9, because $9 + 1 \neq 8$.

solution to an inequality A solution to an inequality is a number that can be used in place of the variable to make the inequality true.

For example, 5 is a solution to the inequality $c < 10$, because it is true that $5 < 10$. Some other solutions to this inequality are 9.9, 0, and -4.

speed Speed is one way to describe how fast something is moving. Speed tells how much distance the object travels in a certain amount of time.

For example, Tyler walks at a speed of 4 miles per hour. Priya walks at a speed of 5 miles per hour. Priya walks faster than Tyler, because she travels more distance in the same amount of time.

spread The spread of a set of numerical data tells how far apart the values are.

For example, the dot plots show that the travel times for students in South Africa are more spread out than for New Zealand.

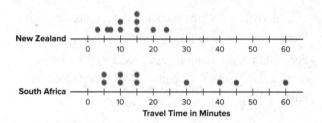

squared We use the word *squared* to mean "to the second power." This is because a square with side length s has an area of $s \cdot s$, or s^2.

statistical question A statistical question can be answered by collecting data that has variability. Here are some examples of statistical questions:

- Who is the most popular musical artist at your school?

- When do students in your class typically eat dinner?

- Which classroom in your school has the most books?

surface area The surface area of a polyhedron is the number of square units that covers all the faces of the polyhedron, without any gaps or overlaps.

For example, if the faces of a cube each have an area of 9 cm², then the surface area of the cube is 6 · 9, or 54 cm².

T

table A table organizes information into horizontal rows and vertical columns. The first row or column usually tells what the numbers represent.

For example, here is a table showing the tail lengths of three different pets. This table has four rows and two columns.

Pet	Tail Length (inches)
Dog	22
Cat	12
Mouse	2

tape diagram A tape diagram is a group of rectangles put together to represent a relationship between quantities.

For example, this tape diagram shows a ratio of 30 gallons of yellow paint to 50 gallons of blue paint. If each rectangle were labeled 5, instead of 10, then the same picture could represent the equivalent ratio of 15 gallons of yellow paint to 25 gallons of blue paint.

10	10	10

10	10	10	10	10

term A term is a part of an expression. It can be a single number, a variable, or a number and a variable that are multiplied together. For example, the expression $5x + 18$ has two terms. The first term is $5x$ and the second term is 18.

U

unit price The unit price is the cost for one item or for one unit of measure. For example, if 10 feet of chain link fencing cost $150, then the unit price is 150 ÷ 10, or $15 per foot.

unit rate A unit rate is a rate per 1.

For example, 12 people share 2 pies equally. One unit rate is 6 people per pie, because 12 ÷ 2 = 6. The other unit rate is $\frac{1}{6}$ of a pie per person, because $2 \div 12 = \frac{1}{6}$.

V

variability Variability means having different values.

For example, data set B has more variability than data set A. Data set B has many different values, while data set A has more of the same values.

Data Set A

Data Set B

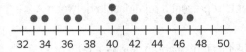

variable A variable is a letter that represents a number. You can choose different numbers for the value of the variable.

For example, in the expression $10 - x$, the variable is x. If the value of x is 3, then $10 - x = 7$, because $10 - 3 = 7$. If the value of x is 6, then $10 - x = 4$, because $10 - 6 = 4$.

vertex (vertices) A vertex is a point where two or more edges meet. When we have more than one vertex, we call them vertices.

The vertices in this polygon are labeled A, B, C, D, and E.

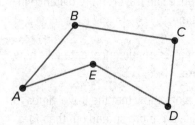

volume Volume is the number of cubic units that fill a three-dimensional region, without any gaps or overlaps.

For example, the volume of this rectangular prism is 60 units^3, because it is composed of 3 layers that are each 20 units^3.

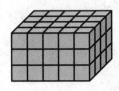
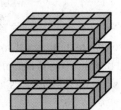

Index

triangles
    area of, 57
    base of, 66–68, 76–77
    height of, 66–68, 76–77
    base-height pairs, 66–68, 76–77
    formula for the area of, 66–68
    opposite vertex, 66–68
    relationship to
        parallelograms, 48
    with fractional lengths, 445

 U

unit price, 187, 201, 286

unit rate, 279
    and equivalent ratios, 286

units, conversion of, 268
units of measurement,
anchoring, 256

 V

variability, 857, 933, 940

variable, 590, 616

variation, *See variability.*

vertex (vertices), 77, 84

vertical axis, *See y-axis.*

volume, 118
    of prisms, 453
    distinguishing between surface
        area and, 117–118

 X

x-axis, 780

x-coordinate, 780

 **Y**

y-axis, 780

y-coordinate, 780